To the late Alice Ann K. ("Bincy") Taylor (1928–2015)

Names: Taylor, Snowden, author. | Dapkus, Mary Jane, author.

Title: Antebellum shelf clock making in Farmington and Unionville villages, Connecticut / by Snowden Taylor and Mary Jane Dapkus.

Description: Columbia, PA : National Association of Watch & Clock Collectors, Inc., [2018] Includes bibliographical references.

Identifiers: LCCN 2018017827 | ISBN 9781944018047

Subjects: LCSH: Clock and watch making--Connecticut--Farmington--History--19th century. | Clock and watch making--Connecticut--Unionville--History--19th century. | Clock and watch makers--Connecticut--Farmington. | Clock and watch makers--Connecticut--Unionville. | Shelf clocks--Connecticut--Farmington--History--19th century. | Shelf clocks--Connecticut--Unionville--History--19th century.

Classification: LCC TS543.U6 T227 2018 | DDC 681.1/13097462--dc23 LC record available at https://lccn.loc.gov/2018017827

Publisher: The National Association of Watch & Clock Collectors, Inc. 514 Poplar Street, Columbia, PA 17512.

Editor: Christiane Odyniec

Copy Editors: Freda Conner and Gillian Radel

Printed in the United States of America.

Cover photo(s)

Seymour, Williams & Porter, Farmington [Unionville], CT, 30-hr. wooden movement shelf clock ca. 1832, in long, carved column and splat-style case with hairy paw feet, and lithograph tablet entitled: "THE BRIDE". The clock's movement (Type 1.151), likely the earliest style produced by the firm, is set into the case back, a practice seen in clocks by early Terry family firms of Plymouth, CT (see Chapter 2). This example, retaining both its original tin plate movement cover on the back, and the original tin dust covers over the pulley apertures, was undoubtedly a premium model. COURTESY OF COLLEEN AND DAVID HOWTZ.

Table of Contents

Preface

The last comprehensive discussion of early nineteenth-century shelf clocks originating in the villages of Farmington and Unionville (both within the Town of Farmington, CT), occurred in the second edition of *Eli Terry and the Connecticut Shelf Clock*, by Kenneth D. Roberts and Snowden Taylor. Now an in-depth update is possible.

Kenneth D. Roberts, in his groundbreaking works: *The Contributions of Joseph Ives to Connecticut Clock Technology 1810-1862*, 1st ed. (1970), and *Eli Terry and the Connecticut Shelf Clock*, 1st ed. (1973), left a remarkable legacy through his painstaking accumulation of deeds, probate records, newspaper items, and local histories, thereby uncovering the structures of the histories of Connecticut's clock making firms. Not only was Roberts a prolific researcher and writer, he was also meticulous about documenting his sources, thus enabling others, including the authors of the present study, to build on his extraordinary work. To him we owe a debt of gratitude.

In the present study, author Snowden Taylor, coauthor with Roberts in the second edition of *Eli Terry and the Connecticut Shelf Clock* (1994), contributed observations and expertise gained over more than five decades in the analysis of period clocks and movements, including data gathered from many examples originating in the villages that are the subject of the present study that were unavailable to Roberts and Taylor in 1994. With a few additions as noted, author Mary Jane Dapkus contributed documentary research and analysis.

Initially, both authors shared the agreeable tasks of sifting through and puzzling over the data, and of writing. However, when author Taylor's vision failed in 2014, author Dapkus assumed the task of finishing the then half-drafted manuscript. Any errors or omissions are hers and hers alone.

Briefly, this work documents the emergence of two distinct clock making "shops" or "branches" in Farmington and Unionville villages. Each possessed distinguishable roots, aspects of which remained discernible in the clocks and movements of the respective shops, and in those of their successor firms. This observation is rendered all the more remarkable not only in light of various dissolutions and reorganizations, but also in light of myriad external factors that buffeted the firms as time wore on, including lawsuits, fires, deaths, and insolvencies.

In addition to the above, the present work connects the clock makers' lives and work with many aspects of early nineteenth-century American history, including, for example, the Panic of 1837, westward migration, industrialization, the temperance movement, and abolitionism. It dispels much of the previous confusion about the firms, their partners, and their chronologies, and contributes considerable dimension to what has previously been written on the subject of Farmington and Unionville clock making. Furthermore, as Roberts and Taylor clearly established in the second edition of *Eli Terry and the Connecticut Shelf Clock*, the present study reaffirms the value of clocks themselves, their labels, and their movements, as primary sources of historical information. We hope that this book will spark new interest in the history of American clock making.

M.J.D., April 27, 2018.

Acknowledgments

The authors thank the many persons who encouraged and assisted us over the course of this project. Our immediate support base consisted of our spouses: Stan Dapkus and the late Alice (Bincy) Taylor. Vital support for this project was also provided by the Unionville Museum, and by the Farmington Historical Society, who generously opened their collections and resources to us.

The following individuals and organizations shared their knowledge, and in many instances, their friendship: Betty Coykendahl; Jay Manewitz; Nancy Taylor; the late Chris Brown; Dr. Charles Leach; Chris H. Bailey; David Ewbank; Colleen and David Howtz; Jim and Tracy Stehlik; Thomas M. Manning; Derek Dudek; Robert Elwell; the late Ward Francillon; the late Jim Lowe; the American Clock & Watch Museum, Bristol, CT; the Illinois Regional Archives Depositories at Springfield and Carbondale, IL; the Farmington Public Library; the Connecticut State Library; the Connecticut Historical Society; and the National Association of Watch & Clock Collectors (NAWCC), its Library and Archives, its Research Committee, and its editorial staff (past and present), including Diana DeLucca, Monica Elbert, Therese Umerlik, and especially Christiane Odyniec. We are very grateful to Gene Gissin for granting us permission to use many of his wonderful photographs; and to Russ Oechsle for his skillful review of the completed manuscript. It is a much more polished product through his efforts.

We also thank the many other individuals who took the time to contribute information and/or photographs for this project, and for author Snowden Taylor's bimonthly column, "Research Activities & News," which appeared in the NAWCC *Bulletin* under his byline for nearly 35 years. Without their efforts, many of the clocks and clock makers described in this study would not have come to light.

Last (but far from least), special thanks are due to Cliff Alderman for his many contributions and suggestions, as well as his keen interest and active encouragement during the course of this project. His gentle prodding provided the crucial impetus that helped transform it into written reality.

CHAPTER 1

Firms of Farmington Village: Cowles, Deming & Camp; Marsh, Gilbert, & Co.; George Marsh & Co.; and Orton, Preston & Co.

Introduction

Published in 1994, the second edition of the groundbreaking book *Eli Terry and the Connecticut Shelf Clock*, by authors Kenneth Roberts and Snowden Taylor[1], established the framework of the history of early nineteenth century clock making in the villages of Farmington and Unionville, both within the Town of Farmington, CT. With the passage of nearly a quarter of a century since its publication, the present authors happened to discover many "new" primary sources and to observe a number of "new" examples of clocks originating in these villages. Gradually, it became clear that a previously untold story was emerging. It is a story as much about the clockmakers and their work as it is about the forces that shaped their lives in the decades leading up to the Civil War—one in which clocks themselves contributed crucial information. It is with special thanks to the late Kenneth Roberts (1916–2000), who was both a keen observer and a meticulous documenter of his sources, that we present the following.

Cowles, Deming & Camp

The impetus for the present work initially sprang from a chance discovery, by author MJD, of 39 pages of surviving court documents pertaining to a single lawsuit, designated by the clerk of the Hartford County County Court as *Seth Cowles vs. George Cowles et al.*[2] For convenience, the authors will refer to this as "the Seth Cowles case." The suit commenced on October 29, 1833, when Seth Cowles addressed the Hartford County County Court on the second Tuesday of November 1833, as follows:

> The Petition of Seth Cowles of Farmington [Connecticut] in said County respectfully showeth—That on the first day of January 1828 the Petitioner was and ever since has been and still is tenant in common with George Cowles, James K Camp, Richard Cowles, and Samuel Deming all of said Farmington in equal shares, of a certain piece or parcel of land situated in said Farmington bounded northerly, easterly and southerly on land of said Samuel Deming, the Farmington canal or a passway, and westerly on Farmington river, containing about two acres, with a Grist mill saw mill, carding-machine shop, and

clock manufactory standing thereon, and that during the whole time aforesaid the parties hereto were and are tenants in common and partners in equal shares of all the rents, profits and proceeds of the said Grist mill, Saw mill, carding machine and clock manufactory, and that during the time aforsaid the said George, James, Richard and Samuel or some or one of them have taken and received to themselves the use and benefit of said estate and of the rents, profits and proceeds of said Mills, machine and manufactory, in greater proportion than the amount of their interest therein, to the amount of three hundred dollars.[2]

The petition continued, praying "your honors as a court of Chancery...to...grant relief." In response, the Court summoned the parties to appear on the specified date. Subsequently, the Court appointed Romeo Lowrey as a Committee [of 1] to investigate. He met with the parties on February 4, 1834, confirming the description of the partnership and what it owned to be as stated above. The amounts and distribution of the firm's earnings, however, remained matters to be determined. Lowrey found the partnership to be in debt, and that Seth Cowles's share of the debt was $16.60.

The Committee's report was presented to the Court "on the 4th Tuesday in March A.D. 1834." According to the report, the "...rent [received] for the Clock manufactory from January 1st 1832 to July 1 1833" was "$196[.]87 ½." And further, "that since the said 1st day of July 1833 nothing has been received by said Respondents for the use of that part of the premises used as a Clock manufactory..."[2]

Seth Cowles filed a "Remonstrance [objection] to the report of the Committee" and a supplementary petition, both of which the Court heard during its March term, 1834. After examining these documents, the Court "set aside" Lowrey's report and appointed George Merrick as a new Committee. The Committee summoned and examined witnesses during July and August 1834.

Merrick's Committee report was addressed to the August 1834 term of the Court. No new information came to light regarding the nature of the partnership. However, the principals all appeared under oath before the Committee, which probed the firm's complex finances more deeply. The Court accepted the Committee report, and based on the report, awarded Seth Cowles $92.14 to complete his share of the proceeds of the partnership, plus $53.92 for court costs, "... and that executions do issue accordingly."[2]

What about the clock manufactory? The following passage extracted from the Committee's report reveals some important details:

> Your Committee however would observe that some time in the year 1831 George Cowles, Richard Cowles, James K. Camp & Samuel Deming, purchased and placed in said carding machine shop, machinery for the manufacture of Clocks to the amount of one hundred Twenty eight Dollars and fifty seven cents, and it has been exclusively used for the manufacture of clocks, and water power belonging & attached to the Common property Sufficient for driving said machinery has been used for that purpose. To the purchase of this machinery at the expense of the Company Seth Cowles expressly dissented and has ever refused all participation in that branch of business. Your Committee therefore did not allow a charge for said machinery as a Debt against the Company...
>
> In relation to the carding machine shop it was proved that to the 15 day of May 1831 it was used exclusively for that purpose for the benefit

of the Company. At that time Seth Cowles leased to the other proprietors for the term of one year Viz right to the room where the carding machine stood to place said clock machinery in and using a part of the water power to drive the same. And the carding machine was placed in another room and improved through the season for the Common benefit. The consideration of the lease was ten Dollars to be paid to Seth Cowles and insurance to be affected by the lessees on the common property, for the Common benefit for the term of one year to the amount of two thousand Dollars. It was proved that the ten Dollars was paid but the insurance was not made. On 1st of January 1832 George Cowles, Richard Cowles, James K. Camp, and Samuel Deming lease[d] to *Marsh & co.* [italics added] the whole of said Carding machine shop including said Clock machinery for the term of three years, with the privilege of using three gates from the floom [sic] to drive said machinery. The consideration of this lease was one hundred and fifty Dollars a year, payable quarterly. In pursuance of this agreement Marsh & co. entered into the occupation of said shop and privilege and have continued to occupy it to the present time [i.e., August 1834]. On said first day of January the carding machine was removed from the shop, and has since continued out of use. To the whole of this arrangement Seth Cowles dissented: and on trial claimed that he was entitled to receive of the respondents a part compensation for the use of his share of this property, and your Committee sustained this claim and allowed rent up to the quarter ending April 1st 1834. It appearing to your Committee that said Camp was the agent of the respondents in holding the security from Marsh & co. & collecting the money thereon due for rent, and that said Camp had actually received the same, your Committee has considered him as responsible therefor, and in the statement of the accounts herin [sic] presented has considered it in his hands. The amount he is made so responsible for amounts on the 29th of October 1833 to the sum of Two hundred thirty five Dollars fifty one cents; and to the 12th of April 1834 to the sum of Three hundred two Dollars Eighty cents.[2]

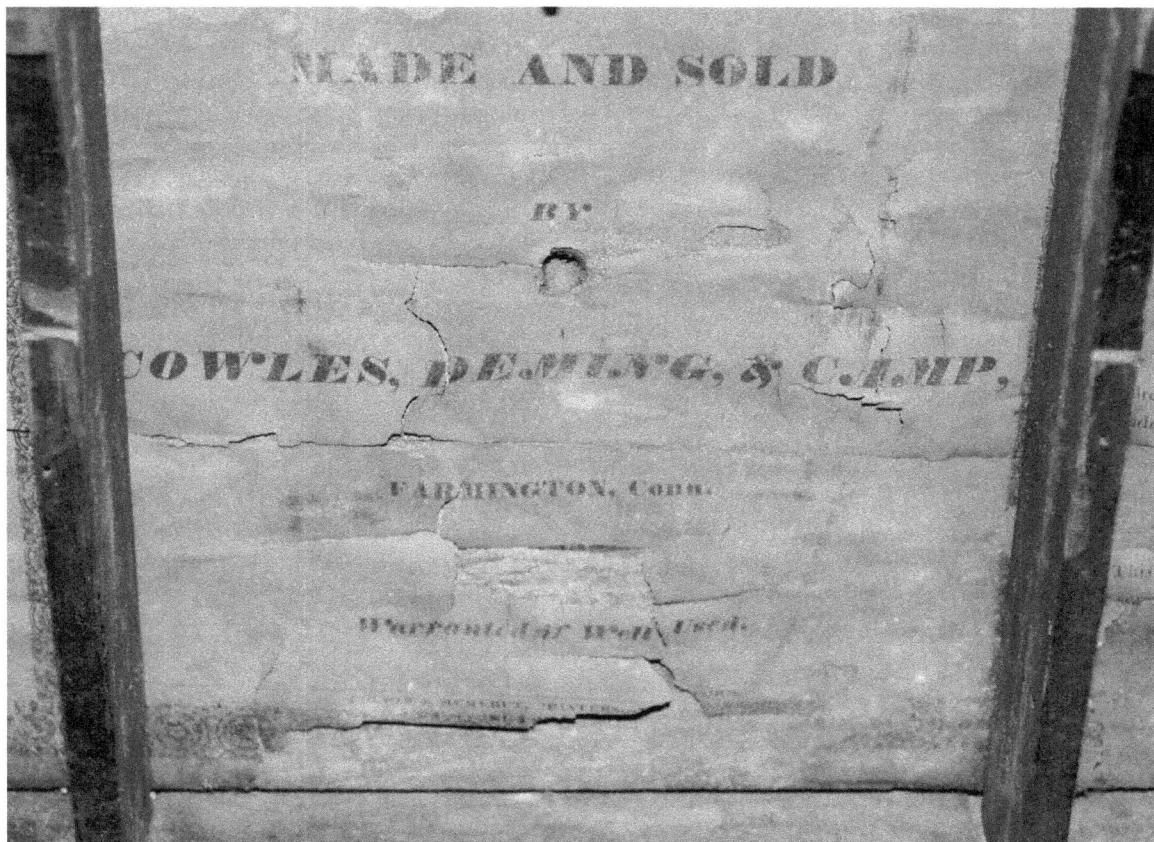

Figure 1. Portion of clock label of Cowles, Deming & Camp, Farmington, Conn. (Courtesy of Cliff Alderman. Photo by R. A. Ruggeri.)

So, although it is unclear from the records whether clock making had already been attempted in the building, beginning on January 1, 1832, "Marsh & co." rented Cowles, Deming & Camp's clock manufactory, and still occupied the facility as of August 1834, paying the rent to James K. Camp.

Previously, in a deed dated November 12, 1827, Samuel Deming, in exchange for $329, transferred a small piece of land to be used as a passway, to:

Seth Cowles, George Cowles, Richard Cowles, Samuel Deming, and James K Camp all of sd Farmington millers oparting [sic] in company by the name of *Cowles, Deming & Camp*. [Italics added. The deed involved]...a certain piece of land in said town, being the

South side of my lot opposite the meeting house in said town, for the sole and exclusive use of a passway from main street [sic] to the mills and appurtenances of the same, belonging to the grantees...and is bounded North on my land; East on main street [sic]; South on land of Catherine Cowles & West on land of the grantees.[3]

This information firmly establishes the existence of the firm of Cowles, Deming & Camp, with the same partners, and a location consistent with that described in the Seth Cowles case. Hence the gristmill, sawmill, carding machine shop, and clock manufactory described therein were all part of the property of the firm of Cowles, Deming & Camp.

A single clock bearing the label of Cowles, Deming & Camp was reported in Reference 1 (page 291), wherein a possible connection with the Farmington firms of Marsh, Gilbert & Co., and George Marsh & Co. was suggested. However, neither photos nor details of the clock, its label, or its movement were available. As luck would have it, as this work was nearing completion, an example was located by Cliff Alderman of the Unionville Museum, who brought it to the authors' attention. The clock, in a large, unusual, half-column-style case (perhaps with later alterations), contained a wooden movement (Type 1.181[18,19]), perhaps original to the clock, made by Chauncey Ives, or the firm of C. & L.C. Ives, of Bristol [Connecticut], inset into the backboard of the case. A photograph of a portion of the clock's label, which reads: "IMPROVED / CLOCKS, / MADE AND SOLD / BY / COWLES, DEMING, & CAMP, / FARMINGTON, Conn...", is shown in Figure 1. Although torn, the printer's line, "FOLSOM & HURLBUT, PRINTERS---[HART]FORD", is legible.

In light of this information, and of the information provided in the Seth Cowles case, clock making seems to have begun in Cowles, Deming & Camp's mill in 1831, shortly *before* the mill's rental by "Marsh & Co." Because additional (hopefully intact) examples of Cowles, Deming & Camp's clocks would appear to hold a key to further unlocking the histories of the clock-making firms of Farmington village, anyone having access to a clock bearing the firm's label is encouraged to contact the NAWCC Research Committee.

A drawing, a portion of which is shown in Figure 2A, said to date to the late 1820s (but more likely the 1830s),[4] shows a mill site matching the description of that of the mill complex belonging to Cowles, Deming & Camp, as referenced in the documents from the Seth Cowles case, and the Samuel Deming deed. Specifically, the mill is bounded on the west by the Farmington River, on the east by the Farmington Canal, and is situated opposite the Congregational Church. However, the drawing is crude and needs some explanation. Note

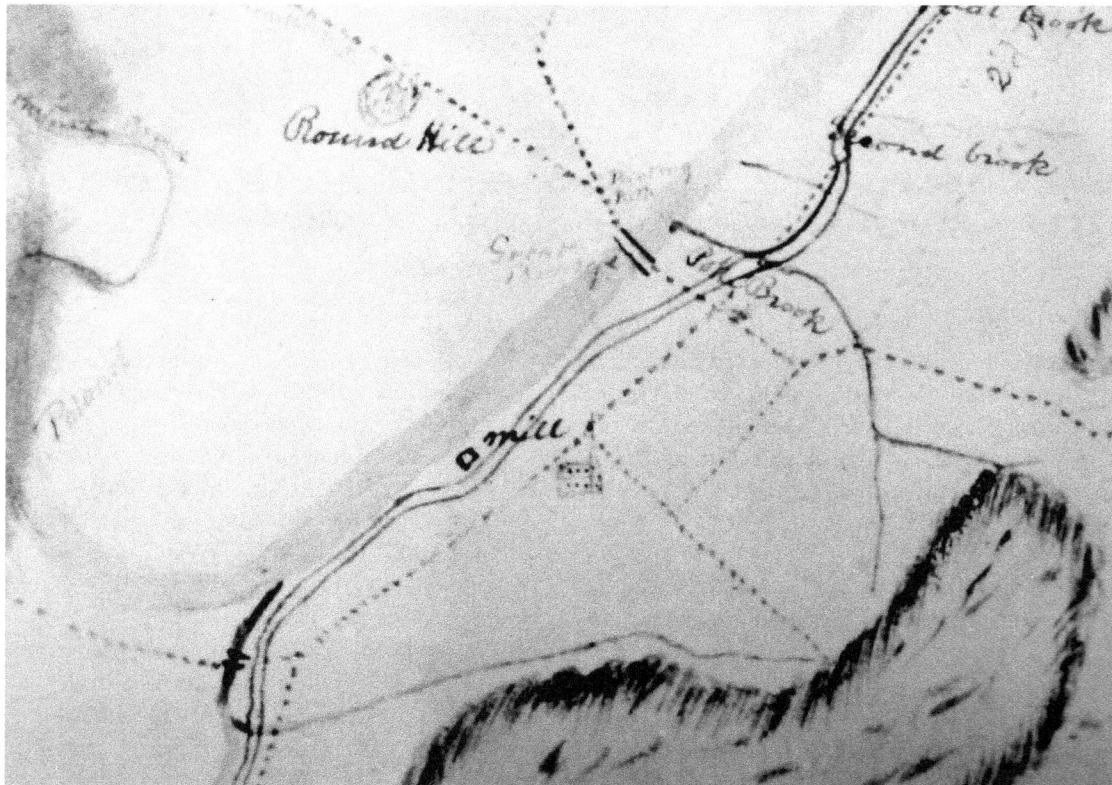

Figure 2A. Crude drawing of vicinity of the mill of Cowles, Deming & Camp. Note "mill," church, and other features. (See text.) (Courtesy of Connecticut State Library.)

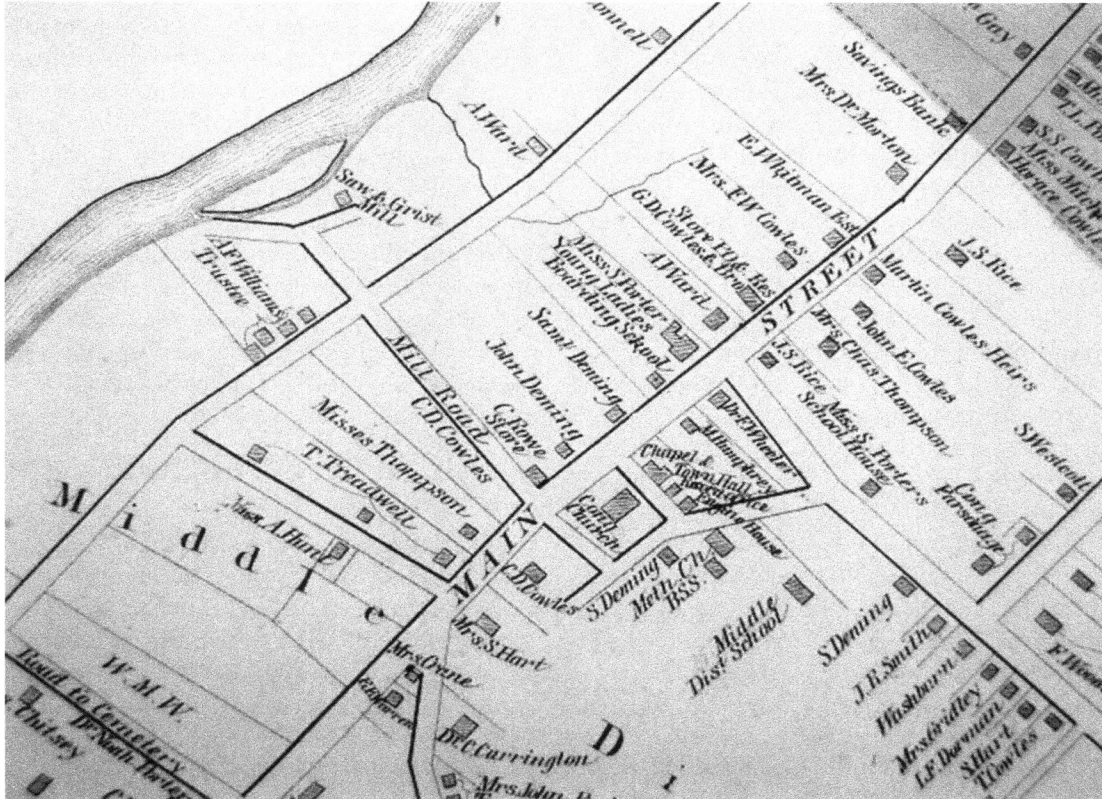

Figure 2B. A portion of Farmington village center in 1869, taken from Baker & Tilden's Atlas *of Hartford and Tolland Counties. Note mill (formerly that of Cowles, Deming & Camp), church, etc. (See text.)*

the "mill" near the center. It lies between a wide gray area to the left (west), which is the Farmington River, and a narrow double line to the right (east), which is the Farmington Canal. Note the sketch of the meetinghouse to the east.

Additional data exist placing the mill site of Cowles, Deming & Camp at this location. Utilizing Baker and Tilden's *Atlas of Hartford and Tolland Counties [Connecticut]* of 1869[5] to leap forward in time by nearly 40 years, a portion of the atlas map of Farmington's Middle School District is shown in Figure 2B, in which the second street running north-south lying east of the sawmill is identified as Main Street. In 1869 the name of Samuel Deming (1798–1884) was still associated with property a little north of the passway, now Mill Road. Note the Congregational Church, located at the intersection of Mill Road with Main Street. A brick building formerly located on Mill Road (leading to the mill), afterward renovated and utilized as a Catholic church, and still later as a meat market, was described in a local history

as having been used during the early nineteenth century as a clock factory,[6] likely in connection with the mill. The Farmington Canal does not appear on Baker and Tilden's *Atlas* map of 1869 because the canal was drained and its haul road used to construct a railroad line during the late 1840s. However, the railroad did not conform exactly to the canal route.[7]

Summarizing our narrative to this point, based on the information provided in the Seth Cowles case documents, together with the information provided on the maps, there can be little doubt that Marsh, Gilbert & Co. and George Marsh & Co. manufactured clocks at Cowles, Deming & Camp's mill about 1832–34. The location of these firms was previously unknown.

What sort of men were the partners in Cowles, Deming & Camp? Because the very best of Farmington's financial, social, and intellectual resources were accessible to them, it is pertinent to provide some genealogical background. One of the partners, George Cowles, was a local druggist. He and two other partners,

Richard Cowles and James K. Camp, had each married daughters of prominent Farmington resident and West Indies trader John Deming.[8, 9] Partner Samuel Deming was John Deming's son.[10] So, except for Seth Cowles, each of the partners in Cowles, Deming & Camp was either a son or a son-in-law of John Deming.

By the turn of the nineteenth century, John Deming, along with his brother and longtime business partner, Chauncey Deming, had become two of the town's wealthiest residents.[7] Richard Cowles's brother Timothy (who plays a prominent role in later chapters) married Chauncey Deming's daughter. Major Timothy Cowles, as he was known, himself a very wealthy man, served as director of a Hartford bank and represented Farmington in the State legislature for several terms. During the 1830s many of the state's prominent business leaders, including Cowles, were allied with the Whig political party. Timothy Cowles held a prestigious State senate seat in 1835, the year in which he also served as a member of the corporation of Yale College in New Haven. Timothy Cowles resided in "...the first house south of the Congregational Church and on the same side of the main street,"[8] in close proximity to the Cowles, Deming & Camp mill complex (Figure 2). As of 1815 this residence was described as the finest in Farmington.[11]

In contrast to the other partners of the firm of Cowles, Deming & Camp, little is known about Seth Cowles, the dissenter relative to the firm's clock-making enterprise. However, it appears likely he was one of five sons of Elijah Cowles of Farmington, all of whom were deeply involved with local banking, financing, and with the building of the Farmington Canal.[7] Another of John Deming's sons-in-law, Gad Cowles, traded in clocks at least occasionally, as we shall see.

Although such views were highly controversial both locally and nationally at the time, at least two members of the firm of Cowles, Deming & Camp, namely, Samuel Deming and James K. Camp, were among a number of local residents for whom religious beliefs connected logically and firmly with the abolition of slavery. During the 1830s and 1840s Deming's home in Farmington served as a station on the Underground Railroad. He was also one of a group of local men instrumental in bringing to Farmington for shelter and education in 1841 the freed African natives who had commandeered the slave ship *Amistad*. Other early abolitionists of Farmington associated with clock making were Timothy Cowles and Cowles's son-in-law, Austin F.

Williams, whose stories will be taken up in greater detail in Chapters 2 and 3. Samuel Deming arranged to temporarily house the *Amistad* mutineers (newly acquitted by the U.S. Supreme Court of any crime in the matter) at his store, located "...at the corner of Main Street and Mill Lane."[7] Based on this description, Deming's store lay in close proximity to the site of Cowles, Deming & Camp's mill. Although clock production at the mill site in Farmington village had ceased by the time the Africans arrived there in 1841, the debate over abolition received much local attention throughout the 1830s and 1840s. We will touch upon this historically important matter again in a later chapter.

Marsh, Gilbert & Co., and George Marsh & Co.

The principal partner in Marsh, Gilbert & Co. and George Marsh & Co., firms that manufactured clocks at Cowles, Deming & Camp's mill complex, was George Marsh. He was born at Northfield, within the town of Litchfield, CT, on September 8, 1794, to parents James and Ursula (Hayden) Marsh. On August 14, 1822, Marsh married Caroline Gilbert,[12] the daughter of James and Abigail (Kenney) Gilbert. George Marsh's business partner, William Lewis Gilbert, was born at Northfield on December 3, 1806. William was Caroline (Gilbert) Marsh's younger brother.[13]

According to the information provided in the extant account books of clockmaker Samuel Terry (now in the collection of the American Clock & Watch Museum), George Marsh began working for Terry at Plymouth in 1827, both Terry and Marsh moving with their families to Bristol in 1828, when Samuel Terry set up a new factory there. Seeing much potential in the expanding clock industry, Marsh and his brother-in-law Gilbert purchased the clock factory formerly belonging to the firm of Elias Roberts & Co., adjacent to land of clockmaker Chauncey Jerome at Bristol, in December 1828. Indeed, Marsh and Gilbert are thought to have been the two men Jerome claimed made their own tools and clock parts to set themselves up in business in another town while being paid to work for him (see Reference 1, pp. 289–291). Based on the information provided in the Seth Cowles case, by January 1832, Marsh and Gilbert had in fact set up their own clock business in Farmington.

Marsh and Gilbert remained in the clock-making business at Farmington only briefly. William L. Gilbert returned to Bristol about 1834, where he became involved with John Birge in the production of shelf clocks with Joseph Ives's 8-day strap brass movements, and with other ventures,[1] marrying Clarinda K. Hine (1815–1874) of Washington in 1835. In 1840 William L. Gilbert moved to Winsted, and in 1841 he purchased much of the former Riley Whiting clock-making facility at that village, which lay within the town of Winchester. There Gilbert began producing 30-hour weight-driven brass movement clocks and pursued a most successful clock-making career. A very wealthy man, he died on June 29, 1890, in Cedardale, near Oshawa, ON, while on a visit to Canada.[13]

Like William L. Gilbert, George Marsh soon left Farmington, but instead headed westward to Ohio. From there he traveled about purchasing inexpensive government land in various counties in Ohio, southern Illinois, and Indiana, for development and speculation. About 1834 he also commenced wood clock making in the town of Piqua, then slightly later, near Dayton, OH. In addition to his many ventures, the town of Marshfield, OH, which he helped establish, and where he died on March 20, 1862, was named after him.[12] Although the two men were no longer copartners, George Marsh maintained personal, business, and clock-making connections with William L. Gilbert through the late 1850s and beyond.

As of the writing of Reference 1, it was not known whether Marsh, Gilbert & Co. or George Marsh & Co., both of Farmington, was the earlier clock-making firm. Nor were the additional partner(s) hidden in the "& Co." of either firm known (Reference 1, pp. 291; 294–297). Surviving documents related to two pertinent lawsuits discovered by author MJD shed new light on these and other matters. A writ of attachment for the first suit[14] states, in part:

> You are hereby commanded to attach to the value of One Hundred & Fifty Dollars the goods and estate of George Marsh of State of Ohio & Cornelius R. Williams of sd Farmington late partners under the name and firm of George Marsh & Co...to answer unto William G. Bates... Town of Westfield... Massachusetts.[14]

So Cornelius R. Williams was the "& Co." of George Marsh & Co., and William L. Gilbert was not involved. (More information on Williams will be presented below.)

William G. Bates commenced his suit in 1842, but legal procedures were often initiated long after the disputed events took place. From additional documents

pertaining to the suit, Bates, apparently an attorney who had provided legal services to George Marsh & Co. in one or more lawsuits brought in Massachusetts courts, eventually won $198.09 in damages and $9.49 in costs.[14] However, no clues regarding the starting and ending dates for the firm of George Marsh & Co. were provided in the records.

In the second court case,[15] a promissory note for $625 dated January 19, 1833, made out to "Marshall and Gilbert in Co.," was signed by Harlow Wilcox of Granby. The justice of the peace who prepared the writ of attachment initiating the case partially caught the error and named the plaintiffs as George Marsh, William L. Gilbert, and Cornelius R. Williams, but the court clerks labeled all the documents *George Marsh & Co. vs. Harlow Wilcox*. Marsh, Gilbert & Co. won the case in 1835, receiving $703.12 in damages and $28.22 in costs.[15] This series of events constitutes a possible, but arguably weak case that Marsh, Gilbert & Co., not George Marsh & Co., was the earlier firm. More on this dating issue later.

In addition to the above, an interesting written agreement in the form of a note was located. This document stated: "Due George Marsh & Co. or order on demand Twenty four hundred Strike pinions for clocks[;] value recd. Farmington July 1st 1835 [signed] Elisha Manross."[16] Manross, a Bristol resident, is known to have provided turned clock parts to the successors of Marsh's firms, first Orton, Preston & Co., and then Williams, Orton, Prestons & Co.,[17] but such transactions between Manross and George Marsh & Co. were previously unknown. The "strike pinions" referred to in the note were probably hammer arbors. The authors do not know what "value" Manross received in return for the strike pinions, or whether Manross had been slow to fulfill his part of the agreement.

At this point it is useful to examine some examples of clocks and movements of the two firms, Marsh, Gilbert & Co. (perhaps the earlier), and George Marsh & Co., both of Farmington village. As set forth in Reference 1 (pp. 291 and 294–297), the present authors agree that the 30-hr. and 8-day wooden movements of both Marsh, Gilbert & Co. and George Marsh & Co. of Farmington are often essentially indistinguishable from each other. However, alarm movements provided opportunities for differences. In their alarm movements, both firms used downward-extended plates, originated by Jeromes & Darrow, with two large holes horizontally arranged in the lower left region of the alarm front

Figure 3A. Marsh, Gilbert & Co., Farmington, CT, clock, 30-hr. with half-column and splat case.

Figure 3B. Label of clock of Figure 3A. Printer's line reads, "JOSEPH HURLBUT, PRINTER- HARTFORD, CONN".

plate. It seems that in Marsh, Gilbert & Co. plates, the left hole is slightly smaller than the right, whereas in George Marsh & Co. movements the two holes are of equal size. Jeromes & Darrow movements, both non-alarm and alarm, are barely distinguishable from these movements. Author ST's updated (but unpublished) movement identification numbers will be used, and a chart linking the current numbers to earlier numbers[18] will be included as Appendix A. Both Marsh firms seem to have used half-column and splat-style cases as were typical of the times. However, it has been suggested that, because many of the cases featured fully carved columns and splats, and many had 8-day wood movements, they were intended for a relatively high-end market.[12]

Figures 3A–3C show a typical Marsh, Gilbert & Co. Farmington clock with a Terry-style Type 7.115[18, 19] 30-hr. wood, weight movement, made by the firm. Its printer's line (Figure 3B) reads: "JOSEPH HURLBUT,

PRINTER - HARTFORD, CONN." For comparison purposes with Figure 3C, a loose Type 7.115[18, 19] 30-hr. movement with an alarm mechanism (albeit missing its escape wheel) is pictured as Figure 4.

Figures 5A–5E show an 8-day Marsh, Gilbert & Co. clock with Type 4.213[18, 20] movement made by the firm. Note the carved eagle splat and carved columns. A close-up of a portion of the label, some-what damaged (Figure 5D), shows the printer's line, "FOLSOM & HU[RLB]UT, PRINTERS, [HART]FORD[, CONN.]" Figure 5E presents the Terry-style 8-day, Type 4.213 movement made by the firm. Figures 6A–6C picture another example, with an 8-day, Type 4.213[18, 20] movement with alarm, also made by Marsh, Gilbert & Co. The tablet may be a lithograph. The printer's line, seen in Figure 6B, reads: "FOLSOM & HURLBUT, PRINTERS - HARTFORD, CONN.", confirming the printer's line in the clock shown in

Figure 3C. Terry-style, 30-hr. wood, weight, movement, Type 7.115, made by the firm. (Figures 3A-3C courtesy of American Clock & Watch Museum.)

Figure 4. Loose alarm movement, escape wheel missing, Type 7.115, for comparison. (Courtesy of Chris Brown. Photo by Ward Francillon.)

Figure 5A. March, Gilbert & Co., 8-day, clock with carved columns and eagle splat.

Figure 5B. Clock of Figure 5A, interior view. (Figures 5A-5B photos by Derek Dudek.)

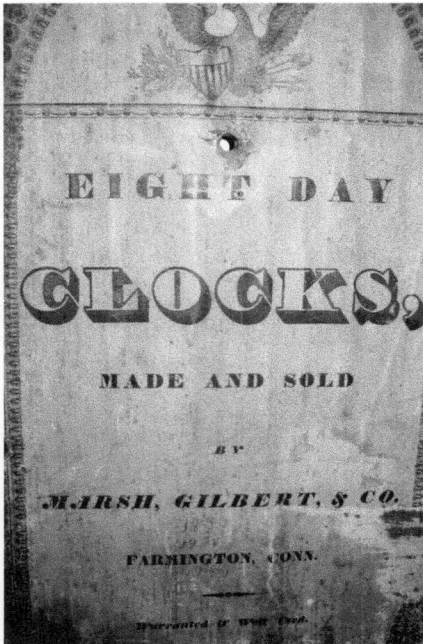

Figure 5C. Label of clock of Figures 5A-5B.

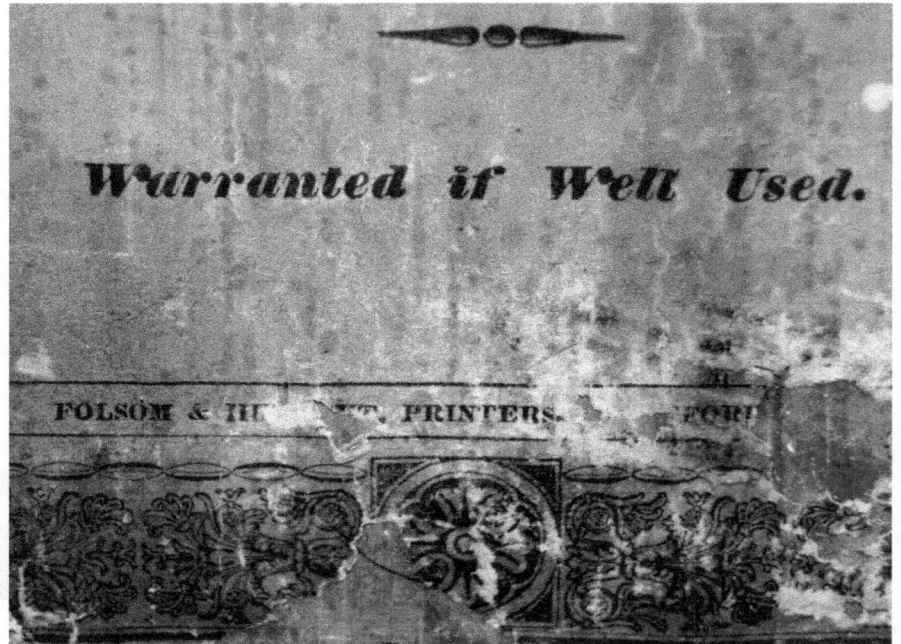

Figure 5D. Enlargement of label of Figure 5C showing printer's line, "FOLSOM & HU[RLB]UT, PRINTERS, [HART]FORD[, CONN.]".

Figure 5E. Marsh, Gilbert & Co. 8-day wood movement, Type 4.213, made by the firm. (Figures 5C–5E photos by MJD.)

Figure 6A. Another Marsh, Gilbert & Co. 8-day carved column and splat-style clock. Tablet may be a lithograph.

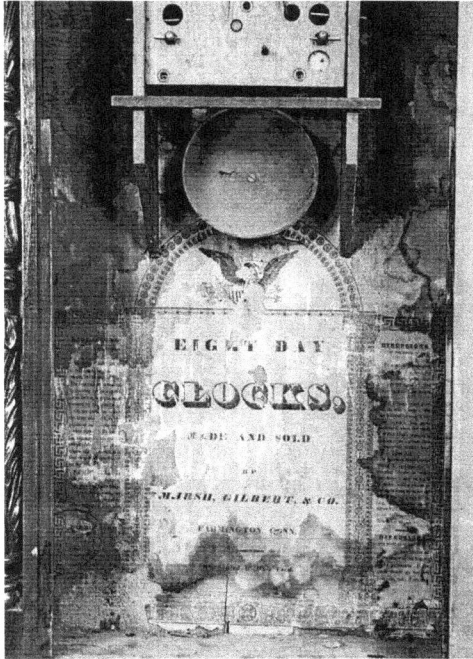

Figure 6B. Label of clock of Figure 6A, again with Folsum & Hurlbut printer's line, confirming that of Figure 5D.

Figure 6C. Type 4.213 alarm movement of clock of Figure 6A. Note the alarm hammer has been broken off and taped to the front plate. Bushings on wind arbors not original. (Figures 6A-6C courtesy of Chris Brown. Photos by Gene Gissen.)

Figure 7. Loose alarm movement, winding arbor bushings not original, Type 4.213, for comparison. (Courtesy of Coleman Bynum.)

Figure 8A. Probable Marsh, Gilbert & Co. double-decker clock with stenciled columns and splat.

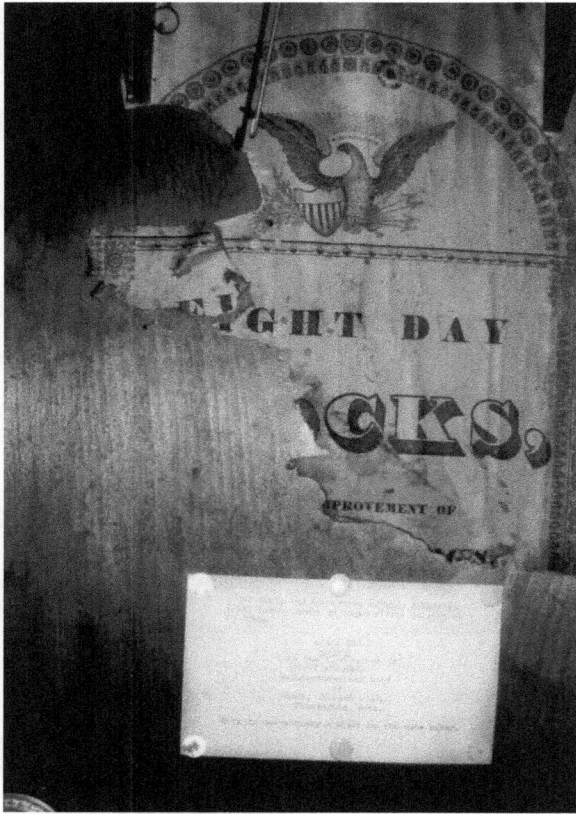

Figure 8B. Badly torn label of clock of Figure 8A.

Figure 8C. Movement of clock of Figure 8A, probably started out to be a Type 4.213 movement, but workman took a short plate and put it into the jig for an extended alarm movement, like that of Figure 7; consequently, the clock had to be finished as a non-alarm. (Figures 8A-8C courtesy of Connecticut State Library.)

Figure 9A. Marsh, Gilbert & Co., 8-day clock with carved columns and eagle splat.

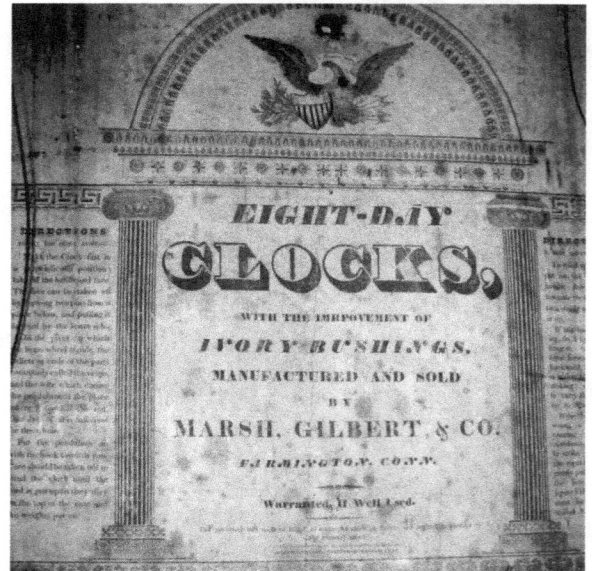

Figure 9B. Label of clock of Figure 9A. Note words, "WITH THE IMPROVEMENT OF / IVORY BUSHINGS". Such bushings, actually bone, are not uncommon on Marsh, Gilbert & Co. clocks, but the firm did not initiate them. Look again at the torn label of Figure 8B, as some of the torn-through words can now be interpreted. The printer's line reads: "JOSEPH HURLBUT, PRINTER - HARTFORD, CONN".

Figure 9C. Terry-style, 8-day, wood, weight, movement, Type 4.212, made by the firm. Note repair work on the wooden block holding the escape cock. Also, the bushing on the time-side winding arbor is not original. (Figures 9A-9C courtesy of Connecticut State Library.)

Figures 5A–5C. Figure 6C shows that the alarm hammer has been broken off and taped to the plate. The bushings on the great wheel arbors are not original. For comparison, a loose, unbroken 8-day Type 4.213[18, 20] alarm movement is shown in Figure 7. It has been rebushed. Also compare these two alarm movements to the 30-hr. alarm movement, Type 7.115, of Figure 4.

Another 8-day clock, probably by Marsh, Gilbert & Co., with a Type "not quite" 4.213[18, 20] 8-day, "not quite alarm" movement, is pictured in Figures 8A–8C. The clock features a stenciled double-decker case. Figure 8B shows the label, badly torn. It is of particular interest in that the movement, shown in Figure 8C, reveals a production error. It would appear that a workman picked up a short non-alarm plate, but put it into a drilling jig for an alarm movement, producing the double hole at lower right and the displaced center lower pillar.

Obviously, it had to be finished as a non-alarm movement, and they used it anyway. (Compare it to the 8-day alarm movement of Figure 7.)

Figures 9A–9C present a Marsh, Gilbert & Co. clock with an 8-day Type 4.212[18, 20] movement (hopefully original to the clock), also made by the firm. Note on the label (Figure 9B), the words "WITH THE IMPROVEMENT OF / IVORY BUSHINGS", which helps interpret the torn-away parts of the label of the previous example, Figure 8B. "Ivory" (actually bone) bushings were not uncommon on Marsh, Gilbert & Co. clocks, although the firm was not the first to use them. Reinforced bushings undoubtedly served as a marketing tool, with which customers could be convinced to pay a little more for clocks having such features that seemed to enhance their durability.

As in the clock shown in Figures 3A-3C, the printer's line reads, "JOSEPH HURLBUT, PRINTER - HARTFORD, CONN." Note also that the movement in Figure 9C has had repairs to the wooden block supporting the escape wheel bridge and that the bushing on the time side winding arbor is not original.

Many clock firms enlarged their offerings by buying-in clocks made by other firms. The next two clocks illustrate that practice. Both are 8-day triple deckers with strap brass movements and labels stating, "MANUFACTURED BY / BIRGE & IVES [1831 - 1833[21]] / FOR / MARSH, GILBERT & CO. / FARMINGTON, CONN." but they have different Birge & Ives movements. The first clock, Figures 10A–10C, has a Type 4.112[18,22] movement, and the second, Figures 11A–11C, has a Type 4.12[18,22] movement. The printer's line for each is, "P. CANFIELD, PRINTER, HARTFORD."

A clock with a Terry-style, Type 7.115 30-hr. George Marsh & Co. wood movement is displayed in Figures 12A–12C. Its movement has a normal upper right pillar (Figure 12C), and its tablet (Figure 12A), apparently a lithograph, is titled *The Soldier's Adieu*. Note that in Figure 12B, the label states, "WITH IVORY BUSHINGS". On the upper part of the label is a sales label stating, "P. HUMBERT DROZ, / DEALER IN ALL KINDS OF / WATCHES, CLOCKS & / JEWELRY. / DETROIT ST. OHIO CITY, O. / [two unreadable lines]". Figure 13 shows a loose 30-hr. Type 7.115[18, 19] alarm movement for comparison. As previously noted, George Marsh & Co. clocks were essentially identical to those of Marsh, Gilbert & Co., and both firms utilized typical case styles of the day.

Figures 14A–14C show a George Marsh & Co. clock with a Type 7.116[18,19] 30-hr. wood, weight movement, made by that firm. Note (Figure 14C) that the upper right movement pillar is displaced to the right. A half-column and eagle splat, 2-door, 8-day George Marsh & Co. clock, with a Type 4.212[18,20] wood movement is displayed in Figures 15A–15C. Note that the feet may be incorrect (Figure 15A) and that the movement's strike side winding arbor (Figure 15C) has been rebushed.[23] All three of the last mentioned George Marsh & Co. clocks above have printer's lines that read: "JOSEPH HURLBUT, PRINTER - HARTFORD", or minor variations thereof.

Label printer data often provide a means of helping to date clocks and clock-making firms. In the case of the Farmington village firms now under consideration, without knowing beforehand which was the earlier of the Marsh firms, such help is limited. However, some label data have been collected and analyzed in the case of the Eli Terry firms (Reference 1, pp. 253–254). Additional useful data are found in D. R. Slaght's "Printers of Hartford [Connecticut] 1825–1860"[24] and in the late Paul V. Heffner's "Printers Directory",[25] a work yet to be completed. In the present case, the authors have found the Terry firm data, mentioned above, to be helpful, as follows.

Figure 10A. Birge & Ives for Marsh, Gilbert & Co., triple-decker shelf clock with 8-day brass movement.

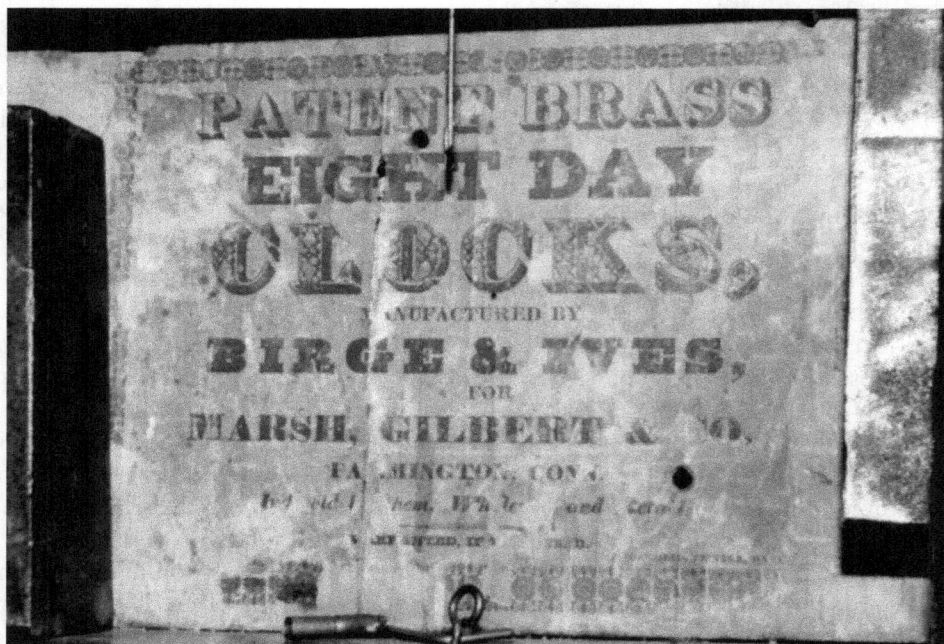

Figure 10B. Label of clock of Figure 10A, stating, in part, "PATENT BRASS / EIGHT DAY / CLOCKS, MANUFACTURED BY / BIRGE & IVES / FOR / MARSH, GILBERT & CO. / FARMINGTON, CONN". Printer's line reads, "P. CANFIELD HARTFORD".

Figure 10C. Strap brass, 8-day, weight movement by Birge & Ives, Type 4.112, from clock of Figure 10A, made by Birge & Ives. (Figures 10A–10C courtesy of Fred Marr. Photos by Bob Edwards.)

Figure 11A. Another Birge & Ives for Marsh, Gilbert & Co., 8-day triple-decker, containing strap brass movement.

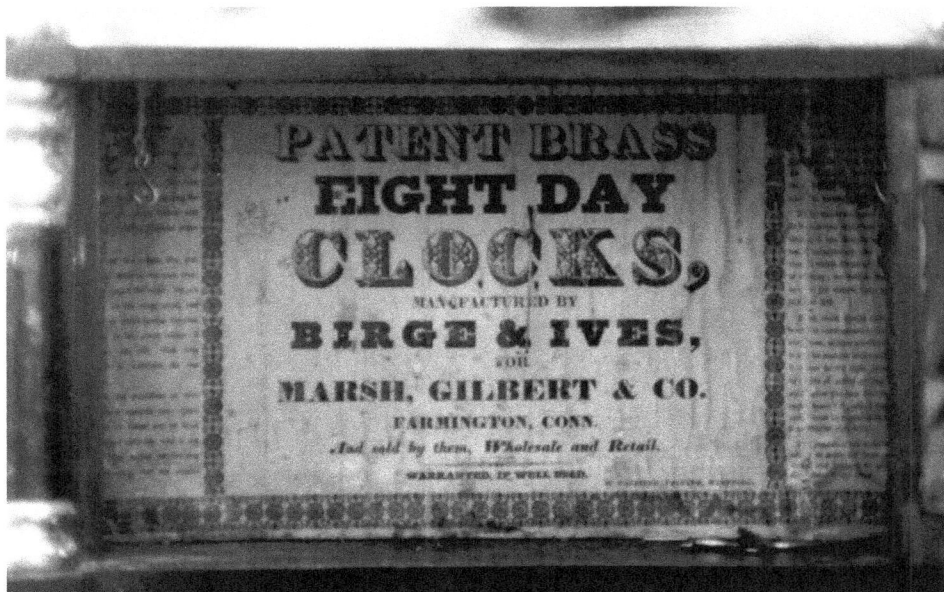

Figure 11B. Label of clock of Figure 11A, wording same as on Figure 10B, including printer's line.

Figure 11C. Strap brass, 8-day, weight movement by Birge & Ives, Type 4.12, from clock of Figure 11A. (Figures 11A–11C photos by Jerome R. Stocking.)

Figure 12A. George Marsh & Co., 30-hr half-column and splat case. Tablet is a lithograph, The Soldier's Adieu.

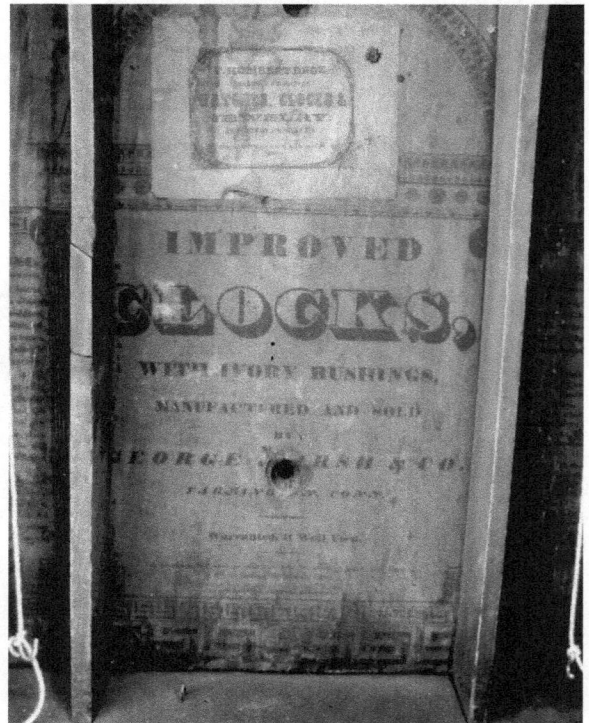

Figure 12B. Label of clock of Figure 12A. Note, "WITH IVORY BUSHINGS". Also note advertising label above, reading, "P. HUMBERT DROZ. / DEALER IN ALL KINDS OF / WATCHES, CLOCKS & / JEWELRY. / DETROIT ST. OHIO CITY, O. / [two unreadable lines]".

Figure 12C. Type 7.115, 30-hr. wood movement of clock of Figure 12A, made by the firm. Note movement is indistinguishable from the Marsh, Gilbert & Co. movement of Figure 3C. (Figures 12A-12C courtesy of American Clock & Watch Museum.)

Figure 13. Loose Type 7.115, 30-hr. wood alarm movement. Compare with Figure 12C. (Photo by Lindy Larson.)

Figure 14A. George Marsh & Co., 30-hr. half-column and splat cased clock.

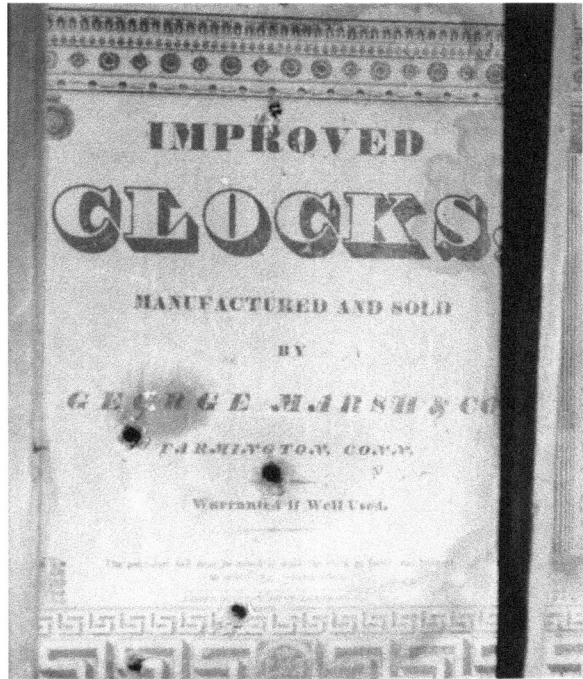

Figure 14B. Label of clock of Figure 14A. Printer's line reads: "JOSEPH HURLBUT, PRINTER - HARTFORD, CONN."

Figure 14C. Terry-type, 30-hr. wood, weight movement, Type 7.116, made by the firm. Note upper right pillar is displaced to the right. (Figures 14A-14C photos by R. K. Hughes.)

Figure 15A. George Marsh & Co., 8-day, smooth column and carved eagle splat, two- door clock.

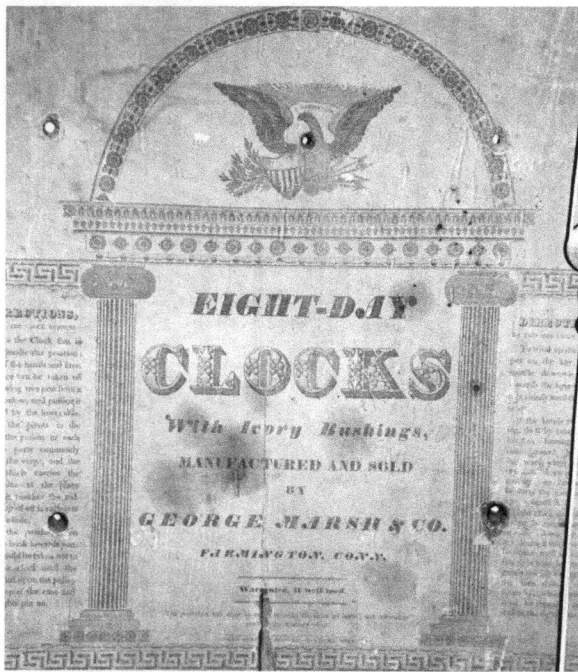

Figure 15B. Label of clock of Figure 15A. Printer's line is, "JOSEPH HURLBUT, PRINTER - HARTFORD, CONN."

Figure 15C. Type 4.212, 8-day, movement of clock of Figures 15A-15B. Note strike side winding arbor bushing is not original. (Figures 15A-15C photos by H. Bryan Rogers.)

Figure 16B. Label of 8-day George Marsh & Co. clock of Figure 16A. Note lack of any town location.

Figure 16A. George Marsh & Co., 8-day beveled-case clock with no town location.

Figure 16C. Type 4.213, 8-day, wood, weight, movement by George Marsh & Co., of clock shown in Figures 16A-16B. (Figures 16A-16C photos by H. Bryan Rogers.)

To summarize the printer data presented above for Marsh, Gilbert & Co., four clocks were discussed in some detail. Two had labels by Folsom & Hurlbut, and two by Joseph Hurlbut. For George Marsh & Co., three clocks were discussed in some detail, and all three had labels by Joseph Hurlbut. For the Terry clock firm data (Reference 1, pp. 253–254), both of our printers are discussed, with dates. Folsom & Hurlbut, dated 1831–32 based on the evidence at hand, was earlier than Joseph Hurlbut by himself, whose dates run from 1832 to 1838 and also from 1840 to 1845 (Hurlbut was a partner in Hurlbut & Williams in 1839). Admittedly, these dates were as applied to the Terry firms, not our two Farmington firms. However, the authors believe that this strengthens our previous suggestion that the firm of Marsh, Gilbert & Co. preceded George Marsh & Co.

The authors agree with Reference 1 (pp. 289, 297, and 300) that the George Marsh, and George Marsh & Co. clocks with labels lacking place names, were probably made in Connecticut for Marsh, some or all in his former Connecticut shop, after he had departed for Ohio in about mid-April 1834. These could have been marketed in either or both states, or perhaps beyond. Figures 16A–16C show a George Marsh & Co. clock in a beveled case, with an 8-day 4.213[18, 20] movement (Figure 16C). Note that the label (Figure 16B) gives no town location. Compare this photo with that of Figure 15B. The next two clocks, each with the label of George Marsh & Co. but having no town name, are housed in similar, but not identical, hollow column cases, and both with Birge-firm 8-day strap brass movements, are shown in Figures 17A–17C and 18A–18C. The first has a carved eagle top (Figure 17A), and a John Birge, or Birge & Ives (ca. 1831–1835) Type 4.12[18, 22] movement (Figure 17C), while the second has a cornice top (Figure 18A) and a Birge, Case & Co. (1833–1835) Type 5.12[18, 22] movement (Figure 18C). Another clock, Figures 19A–19B, has a label that reads, in part, "MANUFACTURED FOR / George Marsh", with no location given. The 8-day brass movement (Figure 19B; previously reported in Reference 17, p. 76, as Figure 12), is no longer thought to be made by S. B. Terry (Reference 17, pp. 76–77). We will see and discuss it again shortly.

As mentioned above, after moving to Ohio, George Marsh set up the firm of Marsh, Williams & Co., first in Piqua (ca. 1834[12]) and then in Dayton (ca. 1834–1835). The "Williams" in both Ohio firms was the same Cornelius R. Williams who had been Marsh's partner in George Marsh & Co. in Connecticut, but Williams did not move to Ohio.[12] It is believed that Marsh, Williams & Co. made no movements. Perhaps Williams maintained an inventory of some George Marsh & Co., and George Marsh clocks, awaiting the development of the new Ohio firm. However, as we shall see, the firms of Orton, Preston & Co., and later, Williams, Orton, Prestons & Co., seem to have sold movements to Marsh, Williams & Co., through Williams. Alternatively, Marsh, Williams & Co. of Piqua and Dayton may have been set up as an Ohio clock assembly plant and branch outlet for the Connecticut firms Orton, Preston & Co. and Williams, Orton, Prestons & Co.

Figure 17A. George Marsh & Co., hollow column, eagle splat clock with no town name.

Figure 17B. Label of clock of Figure 17A with no town name.

Figure 17C. Brass, strap, 8-day movement, Type 4.12, made by John Birge or Birge & Ives, of clock shown in Figures 17A-17B. (Figures 17A-17C photos by Charles Steger.)

Figure 18A. George Marsh & Co. hollow column, cornice top, clock with no town name.

Figure 18C. Brass, strap, 8-day movement, Type 5.12, made by Birge, Case & Co., of clock shown in Figures 18A-18B. (Figures 18A-18C photos by David Morgan.)

Figure 18B. Label of clock of Figure 18A, with no town name.

Figure 19A. Label of unknown case style clock, stating, "EIGHT-DAY / BRASS / CLOCKS / MANUFACTURED FOR George Marsh"; with no town name.

Figure 19B. Eight-day brass movement, formerly thought to be made by S. B. Terry. More on this movement later. (Figures 19A-19B photos by David Morgan.)

Figure 20A. Marsh, Williams & Co., beveled-case clock, 8-day. Compare with Figure 16A.

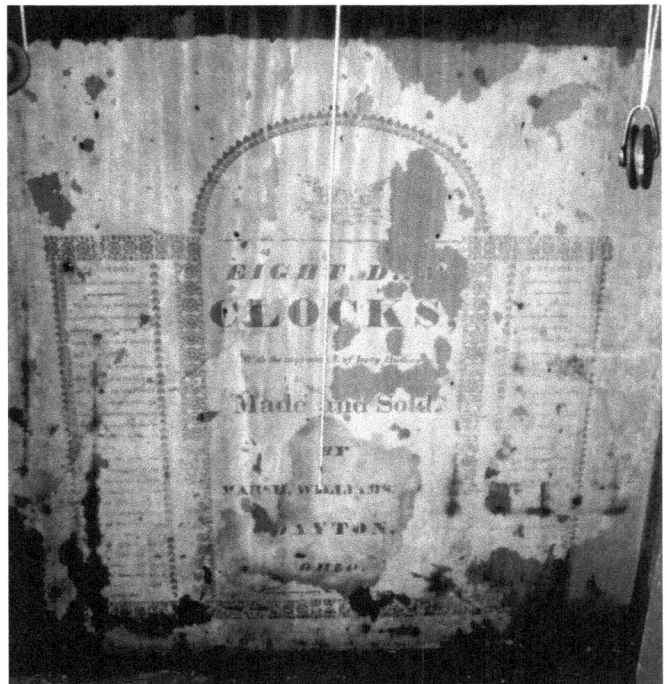

Figure 20B. Label of clock of Figure 20A, which reads, "EIGHT DAY / CLOCKS, / Made and Sold / BY / MARSH WILLIAMS & [CO.] / DAYTON, / OHIO".

We present two examples of Marsh, Williams & Co. clocks. Figures 20A–20C show a Marsh, Williams & Co. beveled-case clock with a Type 4.212[18, 20] 8-day movement. Compare its case to that of Figure 16A. The movement (Figure 20C) has a sheet metal refacing of the count wheel drive pinion block as a repair. A Marsh, Williams & Co. clock with a Terry-style 30-hr. Type 7.153[18, 19] movement, perhaps made by Orton, Preston & Co., is shown in Figures 21A–21C. Note the dial in Figure 21A was probably made in Ohio.[12] See Reference 12 for more on George Marsh in Ohio.

As noted above, Cornelius Robbins Williams was the unnamed "& Co." partner in both the Marsh, Gilbert & Co. and George Marsh & Co. Farmington firms, and the named associate in the Marsh, Williams & Co. Ohio operation. Williams was born in Rocky Hill, CT, on August 7, 1805. About 1831 he joined the Congregational Society in Bristol. His wife, Caroline Hooker, whom he married in Bristol on September 4, 1831, was a descendant of the Reverend Thomas Hooker, seventeenth-century pastor and leader of Connecticut's first settlement.[26, 27] Williams first located in Farmington village about 1833 and then in Unionville.[26, 27] During the period of his involvement with the clock business (between 1830 and 1845[16]), Williams was at first an active member of Farmington's Congregational Church. Then in 1839 he and several others involved in clock making were instrumental in founding Unionville village's first Congregational society. Williams shared antislavery views with Samuel Deming (of the firm of Cowles, Deming & Camp) and with Austin F. Williams.[26] In 1845 or '46, Cornelius R. Williams traveled to Upper Alton, IL, to help wind down the affairs of the clock-making firm of Williams, Orton, Prestons & Co., in which he was a principal partner. He returned to Connecticut in 1847,[16] taking up residence in Terryville, where he died on August 28, 1880.[26] [Note: Cornelius R. Williams is not known to have been related to either Austin F. Williams, a partner in the firm of Seymour, Williams & Porter, or to Anson Williams, a partner in Seymour, Hall & Co., all of whom more will be said in later chapters.]

An important bit of evidence marks a transition of considerable significance to the history of clock making in Farmington. It exists in two versions.[16] The first reads:

I [undoubtedly George Marsh] hereby Certify that I have delivered Orton Preston & Co Five hundred Dollars for which they are to pay me principal & intere[st] at the expiration of their Co. Concern or whe[n] they make a settlement of their business[,] not exc[eeding?] three years[;] should there be no los[s] on their business[,] that is should it pay it[s] way eventually[,] but should sd business prove so bad as to be a losing business then I agree that the said five hundred dollars sho[uld] be used to pay one sixth part of sd loss till the whole be expended[;] should it be neces[sary] to lose the whole for the same G[eorge Marsh?] Farmington April 21st 1834.[16]

The second version, apparently a reply, reads:

Farmington April 21st 1834 Recd of George Marsh Five hundred Dollars to be paid to Sd Geo Marsh at the close of our Business with interest if the Business we are now doing should pay for itself But should there be a loss on said Business then we are to pay him no more than the proportion of three Thousand Dollars is to five hundred Dollars or He is to loose [sic] one sixth of the loss of our Business untill [sic] the whole of the five hundred Dollars is expended.[16]

This seems to be a generous act of concern for the partners of the new firm, proving that Orton, Preston & Co. was truly a successor of George Marsh & Co. Noah Preston, one of the principal partners in the new firm, was the husband of George Marsh's sister. The couple's first child, George Marsh Preston, born in 1834, was named after Marsh. The date in both versions of the document reinforces that found in the Seth Cowles case. Therefore, one can safely say that the transition date from George Marsh & Co. to Orton, Preston & Co. was about mid-April 1834. Many years later the $500 was in fact returned to the former George Marsh & Co.[16]

Orton, Preston & Co.

Reference 1 (pp. 294 and 297) proposed that Orton, Preston & Co. and Williams, Orton, Prestons & Co., in that order, were successors to the Marsh firms, based largely on observed clock case and movement details. No deeds have been found for Orton, Preston & Co., so once again a property lease seems logical. The location was most likely that of the Marsh Farmington firms,

Figure 20C. Type 4.212, 8-day wood movement, possibly made by Orton, Preston & Co. Farmington, CT. Note the repair of the count wheel drive pinion block by installing, seemingly, a piece of sheet metal. (Figures 20A-20C courtesy of American Clock & Watch Museum.)

Figure 21A. Marsh, Williams & Co., 30-hr. two-door, column and splat clock; doors open, revealing the Ohio dial.

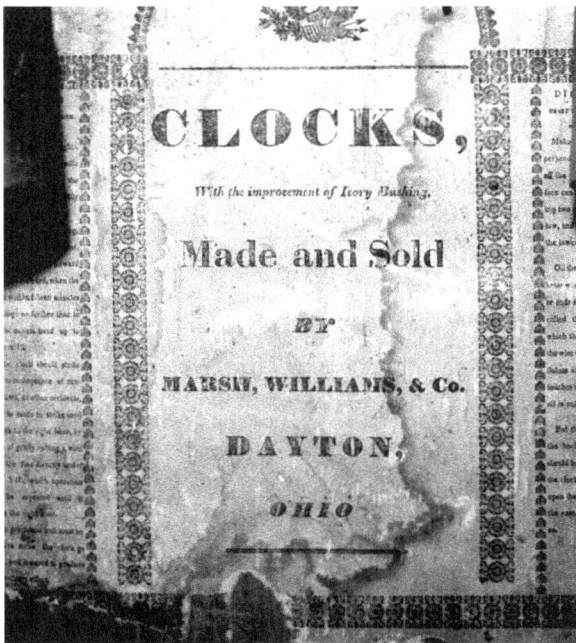

Figure 21B. Label of clock of Figure 21A much like that of Figure 20B, but without words, "EIGHT DAY".

Figure 21C. Terry-style, 30-hr. wood, weight, movement, Type 7.153, perhaps made by Orton, Preston & Co. Farmington, CT. (Figures 21A- 21C photos by Jacques Houser.)

within the Cowles, Deming & Camp mill complex. The starting date for Orton, Preston & Co. was well established by the two surviving documents noted in the last section as "mid-April 1834."

Author MJD has gathered papers related to three different lawsuits pertaining to Orton, Preston & Co. In the first, a writ of attachment reads, in part:

> You are hereby commanded to attach to the value of one hundred dollars the goods or estate of Lucius Parsons and Joel N. Churchill both of said Bristol...and then have [them] to appear before the Hartford County Court...on the 2nd Tuesday of November 1835...to answer unto Noah Preston, Heman Orton, Edwin Hall, Benham Beecher and Henry Tolls, all of Farmington...late dealers in Copartnership under the name & firm of Orton, Preston & Co. In a plea of the Case whereupon Plaintiffs declare... That the Defendants...by a...NOTE...dated the 21st day of October...1834, promised the Plaintiffs, by the name of their said firm to pay... the sum of fifty eight dollars in sixty days...Dated at Plymouth this 15th day of August 1835.[28]

The three last named partners in the firm (Hall, Beecher, and Tolls) were previously unknown. The phrase "late dealers" suggests that Orton, Preston & Co. was no longer in existence on August 15, 1835. We note that October 21, 1834, when Parsons and Churchill gave their note, was a few months after George Marsh is thought to have left for Ohio. The authors have taken a cautious view of the phrase "late dealers" in primary sources such as court records, using it as a clue, but not as definitive proof, of a firm's dissolution (in this instance Orton, Preston & Co.). The outcome of the case is unknown.

The second court case was *Lemuel Whitman vs. Orton, Preston & Co.*[29] In this case, Whitman, a prominent local attorney,[30] brought a civil suit dated June 14, 1841, against the firm to recover book debt owed him in the amount of $70, presumably for legal services. The defendants were named and again referred to as "late dealers in company."[29] Henry Tolls was no longer listed as a partner, but we cannot tell the precise date he left the firm. Similarly, records of this court case provide neither the nature of, nor the date on which, the firm Orton, Preston & Co. incurred the disputed debt, and hence contribute no further information regarding how long the firm remained in business.

An extant letter requesting that an attachment be made yielded valuable information. It is presented here in its entirety, as follows:

> Granby March 23rd 1836
> Mr. [Anson] Bates Esq [of East Granby, Connecticut] Dear Sir I have been out to Farmington but have not settled with Orton, Preston & Company. I came up here for the purpose of getting you to make an attachtment [sic] for them. The firm is Orton, Preston & Co.
>
> The partners are:
>   Heman Orton
>   Noah Preston
>   Edward Hall [sic - perhaps Edwin.]
>   Benham Beecher
>   & Henry Toles [sic - perhaps Tolls.]
>
> Mr. Toles is not in company with them now but was when we sold them the mare. I wish you to make an attachment as soon as you can conveniently - The claim you will recollect is 75$ that was the price of the mare we sold her to them on the 22nd of September 1834 - was to send us the clocks in a week they was to let us have the clocks at 6.25 which would make twelve however I do not know as that will make any difference about the attct. I shall be up here some time Day after to morrow & hope you will have it done Respectfully yours &c Anson Pinney
>
>   You will recollect
>   our firm Griswold & Pinneys
>   Origin P. Griswold
>   Anson Pinney
>   Lester C. Pinney[31]

We learn quite a bit from this letter. Orton, Preston & Co. started by September 22, 1834. Henry Tolls dropped out as a partner prior to March 23, 1836. However, it is doubtful Orton, Preston & Co. still existed at that date.

Three days after the date of Anson Pinney's letter to Mr. Bates, the firm of Griswold & Pinneys initiated its lawsuit against the partners in the firm of Orton, Preston & Co., who were summoned to appear before the County Court in Hartford on the second Tuesday of August 1836. The case file of "Griswold & Pinneys vs. Orton, Preston & Co."[32] revealed that not only was

Table 1. Notes involving Orton, Preston & Co.[16]

Note	Date	Names	Amt. of Note	Remarks
1.	January 4, 1835	Orton , Preston & Co. to Frank Avery	$27.00	
2.	May 7, 1835	Orton, Preston & Co. to Frank Avery	90.00	
3.	November 1, 1835	Orton, Preston & Co. to George Brock	54.00	On the back, in part: "Brock $54.00"
4.	January 13, 1837	Orton, Preston & Co. to Samuel Stin[e] (See remarks.)		"20 Boxes of Good Super[?] Cigars"
5.	July 11, 1837	A. & S. Osborn to Orton, Preston & Co.	100.00	On the back, at one end: "This note was paid with my own funds[signed] N Preston". At the other end: "May 26, 1840/Then received on this/Note/ One Note of Edward Blake/ slee 1.00//Cash 17.25".

Orton, Preston & Co. obligated to deliver to the plaintiffs "Twelve good eight day Clocks without the weights" in exchange for the "Bay Mare" but also that the clocks were to be "sent... to the Plaintiffs at Freeport Alleghany County in the state of Pennsylvania in the course of the then next week...."[32] Furthermore, Griswold and the Pinneys alleged Orton, Preston & Co. had failed to pay for two additional mares purchased prior to September 22, 1834, one of which was likewise to be paid for in twelve "other good eight day Clocks without the weights" delivered at Freeport.[32] The outcome of the case is unknown.

The authors also examined records in the case of George Peck vs. Orton, Preston & Co., but these yielded no useful data regarding the firm's ending date.[33]

A receipt for clocks apparently reaching the outlet of the Farmington Canal in New Haven, to be shipped from there to Burlington, VT, was discovered. It reads:

Received Farmington May 19. 1835. of Cowles & Co for Orton Preston & Co. 2 Boxes Clocks & 1 Weights Marked Geo Brock Care Mayo & Follett. Burlington Vt. to be delivered to the care of S. Trowbridge N Haven [signed] Stiles Bradley.[16]

On the back of the receipt is marked, "Stiles Bradley / Receipt for 2 Boxes Clocks".

Five notes involving Orton, Preston & Co. were discovered, all listed in Table 1.[16] All of the notes were marked "Farmington" except the Stine note, which was marked "Southwick" [probably Massachusetts].

Note 5 in Table 1, A. & S. Osborn to Orton, Preston & Co., is interesting, because (as we shall see) the firm of Orton, Preston & Co. was no longer in existence as of July 11, 1837, when the note was given. Furthermore, an "S. Osborn" "Bronzed" clock case columns, scrolls, and glass, and performed other work for Noah Preston, and also for "Williams & Co.,"[16] presumably referring to Orton, Preston & Co.'s successor firm, Williams, Orton, Prestons & Co. "S. Osborn" may refer to Sheldon Osborn [sometimes "Osborne"] (1800–81), whose name appeared on a few wood movement shelf clock labels originating in Harwinton, CT. It is speculated that "A. Osborn" likely refers to Amos Osborn (1792–1856), also of Harwinton. According to the Harwinton Vital Records, Amos Osborn married clockmaker George Marsh's sister Laura in 1821. Whether Sheldon and Amos Osborn were related to one another is unknown.

An article in author ST's *NAWCC Bulletin* column "Research Activities & News,"[17] summarized in

Reference 1 (pp. 294), provides some relevant information. This article presented, among other things, a copy of a bill sent to "Williams, Orton, Prestons & Co." by druggists Seymour & Dickinson of Hartford, CT, dated December 20, 1837. The bill was included with a 1927 letter written by Edward N. Preston to a clock collector owning a Williams, Orton, Prestons & Co. clock. Mr. Preston, a grandson of Eli Dewey Preston, correctly gave the names of some of the partners of Williams, Orton, Prestons & Co., but included another brother, John Preston.[34] While the authors have seen no direct evidence that John S. Preston Jr. (1808–1868) was ever a partner in the firm, information will be presented later, in Chapter 4, suggesting that this man was in fact connected with Williams, Orton, Prestons & Co. in Unionville village for a brief period, during 1837–1838.

In his letter, Edward N. Preston went on to state (correctly, as it turns out) that Williams, Orton, Prestons & Co. had a "depot" in Alton, IL. The article also summarized the information available at that time for dating of the firms Orton, Preston & Co. and Williams, Orton, Prestons & Co. These dating items, listed as (a) through (d), present data as follows: Item (a) gives dating data on customers of interest in Elisha Manross's sales of turned clock parts. Although every one of the customers' names is slightly different as given, the first three are variations of "Preston Orton Co." (no Williams), dated 1835–1836, and the last four are variations of "Williams Preston's Orton & Co." (all starting with Williams), dated 1836–1841. Item (b) presents the data from the bill mentioned above for "Williams, Orton, Prestons & Co.," namely, the date December 20, 1837. Item (c) gives the dates for letters mentioning "Orton & Preston" in the Eli Terry, Jr. & Co. Letterbook #2,[35] which are all 1839, but refer to clocks sold some years previously. Item (d) indicates dates of 1840 and 1843, which are relevant to Williams, Orton, Prestons & Co., but contribute no information with which to establish either the firm's beginning or ending dates. These data would seem to put the transition from Orton, Preston & Co. to Williams, Orton, Prestons & Co. as "about 1836," with a smooth transition. The first court case considered in this section gave us a "weak" ending date of "prior to August 15, 1835" for Orton, Preston & Co. However, included with the court case records are several notes signed by the firm, with two dating well into 1837. Considering all the above cumulatively, the best we can do for the ending date of Orton, Preston & Co. is 1835–37. Later it will be shown that a better estimate would be "late 1835."

The following is a summary of the various reported types of Orton, Preston & Co. clocks. Figures 22A–22C show an example of an unusual cornice and false hollow column clock, which seems to have been a specialty of Orton, Preston & Co. The label (Figure 22B) seems designed for clocks like this. The movement (Figure 22C) is Terry-style 30-hr. wood, weight Type 7.153[18, 19] made by Orton, Preston & Co. A corniced, modified triple-decker clock, quite unusual, by Orton, Preston & Co., is pictured in Figures 23A–23C. Its label (Figure 23B) is somewhat torn but readable. The movement (Figure 23C) is again 30-hr. Type 7.153[18, 19] but has a replaced all-brass escape wheel bridge. Note the company name, "ORTON, PRESTON & CO. / FARMINGTON. CONN." below the cornice. An Orton, Preston & Co. half-column and splat-style case with an alarm version of the 30-hr. Type 7.153[18, 19] wood movement is shown in Figures 24A–24C. Its condition is very poor; the splat is missing, the tablet replaced, and the label torn. The label (Figure 24B) clearly gives the company name, but the name on the printer's line, with its potentially valuable data, is completely missing. Despite the clock's condition, its movement, Type 7.153[18,19] (shown in Figure 24C) runs. It is interesting to compare the Orton, Preston & Co. 30-hr. movements to those of Marsh, Gilbert & Co., plus George Marsh & Co. 30-hr. movements, excluding the movement in the latter group, which has the upper right plate pillar displaced to the right. One then sees that the remaining principal difference is that the Marsh firms' movements have the "Figure 8" access hole consisting of three small holes of equal size, while in the majority of Orton, Preston & Co. movements, the center holes are somewhat larger than the outer ones.

All Orton, Preston & Co. 8-day clock movements appear to be Type 4.212[18, 20]. Possibly an early example is shown in Figures 25A–25D. The clock (Figure 25A) is a two-door column and splat type, with much of its top (the splat and one capital) missing. The label, shown in Figure 25B, is quite simple and probably an early version. It appears to have no printer's line. The movement (Figure 25C) has a "Figure 8" opening composed of three small, equally proportioned holes, and excessively large brass bushings on the winding arbors. The latter may not be original. Figure 25D shows the movement back plate. A second 8-day clock is shown in Figures 26A–26C. This clock is in even worse shape than the 30-hr. example shown in Figures 24A–24C, lacking all but one corner capital, as well as its dial and lower door

Figure 22A. Orton, Preston & Co., 30-hr. false hollow column and cornice top clock.

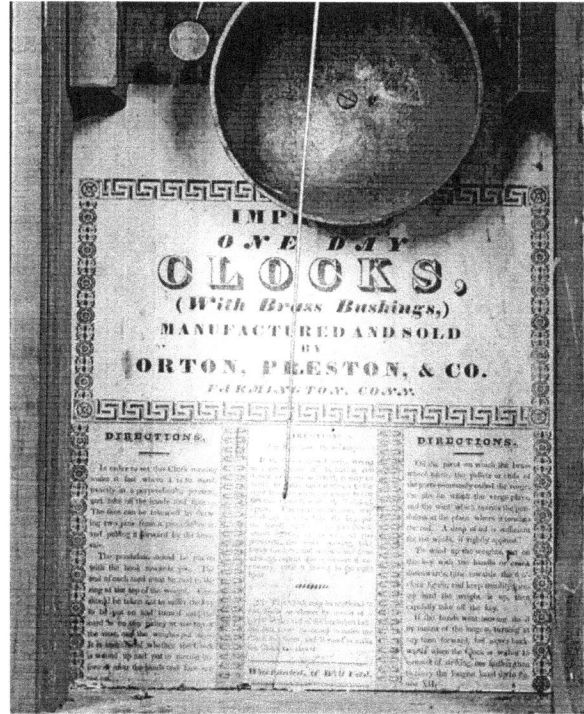

Figure 22B. Label of clock of Figure 22A, seemingly designed for this style clock.

Figure 22C. Type 7.153, 30-hr. wood, weight, movement, made by Orton, Preston & Co. (Figures 22A–22C courtesy of Chris Brown. Photos by Gene Gissen.)

Figure 23A. Orton, Preston & Co., 30-hr. corniced, modified triple-decker clock.

Figure 23B. Label of clock of Figure 23A; although badly torn, all critical material can be read or deduced.

Figure 23C. Type 7.153, 30-hr. wood, weight, movement, mounted in upper part of case. Note words, "ORTON, PRESTON & CO. / FARMINGTON, CONN", below, in cornice. Also note escape wheel bridge has been replaced by a bent metal piece without a wooden block. (Figures 23A-23C photos by Bob & Bernice Elwell.)

Figure 24A. Orton, Preston and Co., half-column and cornice-type case, 30-hr alarm clock. Cornice missing, tablet replaced, and parts of label missing.

Figure 24B. Label of clock of Figure 24A; much of label lost, but "Orton, Preston & Co." readable.

Figure 24C. Type 7.153, 30-hr. wood, weight, alarm movement, made by the firm. It runs! (Figures 24A–24C photos by MJD.)

Figure 25A. Orton, Preston & Co., 8-day, half-column and splat-style case, two door, with most of top ornamentation missing

Figure 25B. Label of clock of Figure 25A; seems early. No printer's line.

Figure 25C. Type 4.212, 8-day wood, weight, movement, made by Orton, Preston & Co. Wind arbor bushings may be replacements.

Figure 25D. Back view of movement in Figure 25C. (Figures 25A–25D photos by MJD.)

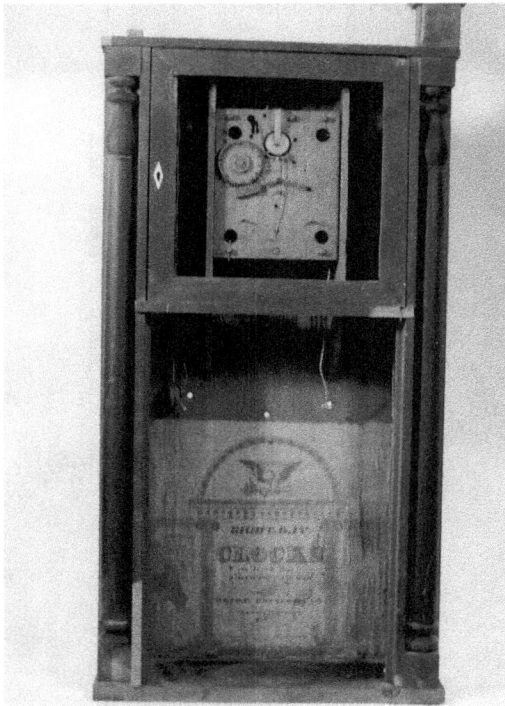

Figure 26A. Orton, Preston & Co., 8-day, wood movement shelf clock in half-column and splat-style case. Most of top, upper glass, and lower door missing.

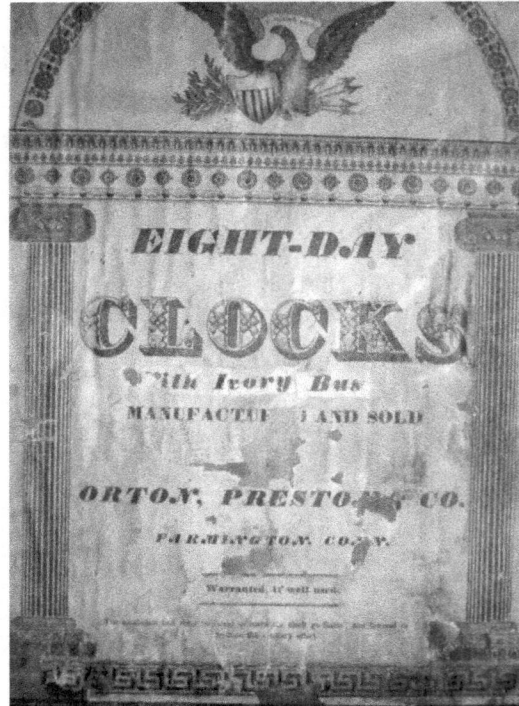

Figure 26B. Label of 8-day Orton, Preston & Co. clock of Figure 26A. Note overpasted strip reading, "With Ivory Bus[hings]". No printer's line.

Figure 26C. Type 4.212, 8-day, wood, weight, movement, made by Orton Preston & Co., of clock shown in Figures 26A-26B. (Figures 26A–26C photos by MJD.)

Figure 27A. Orton, Preston & Co., 8-day, re-stenciled column and cornice case.

Figure 27B. Label of clock of Figure 25A; badly torn. Examination with glass shows no printer's line.

Figure 27C. Type 4.212, 8-day, wood, weight, alarm movement, made by Orton, Preston & Co. However, examination of the "SSS" of the "Figure 8" hole shows that the center hole is slightly larger than the two end holes, perhaps requiring reclassification. (Figures 27A–27C photos by MJD.)

(Figure 26A). Its label (Figure 26B), however, is fairly good. Note that a pasted-on strip reads, "With Ivory Bus[hings]" and that it, too, lacks a printer's line. The movement (Figure 26C) is in fairly good shape. A third 8-day clock is presented in Figures 27A–27C. Note (Figure 27A) that the clock is again a double-door, and it has been re-stenciled. The label (Figure 27B) is in poor shape, and when examined with a glass, no printer's line is apparent. The movement (Figure 27C) is an alarm version with extended plates. Note the center hole of the "Figure 8" opening is slightly larger than the other two holes. This is a new development. Figure 28 is an Orton, Preston & Co. label with a sufficient amount remaining to help fill in the blank areas of the badly torn label of Figure 27B. Actually, it is a slightly later label, and its printer's line reads, "JOSEPH HURLBUT / PRINTER - HARTFORD, CONN." Printer Joseph Hurlbut was discussed above.

Previously unknown partners Edwin (sometimes "Edward") Hall, Benham Beecher, and Henry Tolls (sometimes spelled "Toles") in the firm of Orton, Preston & Co. have been revealed, and we now know that Tolls departed before the end of the firm. An ending date of "late 1835" for Orton, Preston & Co. would not be inconsistent with court records and deeds collected relating to the presumed successor firm to Orton, Preston & Co., namely, Williams, Orton, Prestons & Co., of which more will be said in a later chapter.

Clock casemaker Heman H. Orton, one of the principal partners in Orton, Preston & Co., was born on March 30, 1804, in Litchfield Township, Herkimer County, NY, the son of Heman and Sarah (Hull) Orton.[10] Prior to arriving in Connecticut, it is thought that Orton had spent some time in Vermont,[1] although the present authors have seen no data that either confirm or refute this. However, Heman H. Orton settled in Farmington (Unionville) sometime prior to April 1, 1833, when, according to Volume 2 of the Town Records, he was admitted as an elector (voter). His first wife was Mary Ann. A daughter Helen was born to Heman and Mary Ann in Connecticut about 1836. Mrs. Mary Ann Orton died on January 23, 1838, at the age of 29. Sometime after her death, Heman Orton married (2nd) Jennet, who died on May 1, 1849, at the age of 33.[36] By the time of the 1850 U.S. Census, Heman H. Orton had married again, to Hannah, a native of New York State. As of that date, Orton and his family were still living in Farmington (Unionville), where and when his occupation was still listed as "Clock

Maker." Heman H. Orton died in Unionville on May 29, 1856, at the age of 51.[10, 37] He was buried in the Old Farmington Cemetery alongside his three wives, two of whom had predeceased him.[36]

As previously mentioned, the "Preston" in Orton, Preston & Co. was Noah Preston (1800–1845). Noah's brothers Eli Dewey Preston (1804–1887) and Gardner [sometimes "Garner"] Preston (1806–1869), were also connected with clock making. The three were sons of John Stiles and Aurelia (Dewey) Preston of Harwinton.[34] We will meet Eli D. Preston in a later chapter, as a partner in the firm of Williams, Orton, Prestons & Co. In 1823, as mentioned above, Noah Preston married Lucy Marsh, daughter of James and Ursula (Hayden) Marsh of Northfield, CT[34], and a sister of clockmaker George Marsh (Reference 1, p. 297). The couple's son, George Marsh Preston, was born at Harwinton in 1834 and died at Plymouth in 1869.[34]

Little is known about the other partners in the firm of Orton, Preston & Co. Henry Tolls seems to have been born in Southington in 1808, where he married Amelia Charlotte Hitchcock on February 5, 1834. It is interesting that, according to the aforementioned Farmington Town Records, Henry Tolls was admitted as an elector on April 1, 1833, "By Certificate...from Bristol" as were William L. Gilbert and Cornelius R. Williams, the same day on which George Marsh was also admitted as an elector "by certificate" from Litchfield, suggesting that the men arrived in Farmington together as a group, and that Tolls may have worked with Marsh, Gilbert, and Williams in Bristol. On August 13, 1844, sometime after his first wife's death, Tolls married Amelia's sister, Harriet.[38] At the time of the U.S. Census of 1850, the Tollses were living in Southington, where Henry pursued the trade of box maker. Orton, Preston & Co. partner Edwin Hall may have married a woman named Martha in Alton, IL, in 1834.[10] Benham Beecher was living in Cheshire at the time of his marriage to Nancy Doolittle of Watertown on October 23, 1828.[38] Beecher was born about 1805 and was living in Litchfield, Barkhamsted, and Farmington (all in Connecticut) at the times of the U.S. Censuses of 1830, 1840, and 1850, respectively. As of 1850 his occupation was that of cabinetmaker.

Important information provided in a deed reveals a connection between the Farmington firm of Cowles, Deming & Camp, the clock-making firm of Orton, Preston & Co., and also the latter firm's successor, Williams, Orton, Prestons & Co. On May 6, 1842,

Cornelius R. Williams, Noah Preston, and Eli D. Preston, all of Farmington, mortgaged a 20-acre parcel (their industrial site), located in the village of Unionville, to a long list of creditors, including Cowles, Deming & Camp, to secure their debts.[39] The deed was conditioned that debts owed to the listed creditors, now mortgage holders, would be paid via six-month notes signed on the date of the deed. Clearly, the firm of Williams, Orton, Prestons & Co. was facing bankruptcy. Heman Orton, although a member of both Orton, Preston & Co. and Williams, Orton, Prestons & Co., was not, at the date of this deed, an owner of company land, which is likely the reason he was not mentioned in the deed.

For each debt, the deed listed the name of the creditor, the amount of the note, and the signer of the note. The firm of Williams, Orton, Prestons & Co. signed the first nine notes. Noah Preston signed one note covering a debt owed to James K. Camp, who was, as stated earlier, a partner in Cowles, Deming & Camp and also known to have collected rents for the firm. The next two notes listed included one in the amount of $38.29 owed to James K. Camp, and one in the amount of $42.68 owed to Cowles, Deming & Camp. Both of these were signed by the firm of Orton, Preston & Co.[39]

The deed offers no further explanations, but the authors suggest that the fact that two notes were signed by Orton, Preston & Co. implies this firm was recognized as a predecessor to Williams, Orton, Prestons & Co., and hence responsible for debts incurred during its tenure. Furthermore, the fact that the note for the only debt owed to Cowles, Deming & Camp was signed by Orton, Preston & Co. implies, but of course does not definitively prove, that Orton, Preston & Co. had started out renting a clock factory from Cowles, Deming & Camp of Farmington village.

Having thus introduced our first group of clockmakers and some of their relationships with one another, we will now shift our attention away from Farmington village, to the relatively uninhabited northwest corner of the town of Farmington, soon to become known as Unionville. Not only had the vicinity recently become a hub of canal transportation, but new and greater sources of water power were also being developed there, for the specific purpose of attracting manufacturers. The emergence of clock making there will be the subject of Chapter 2.

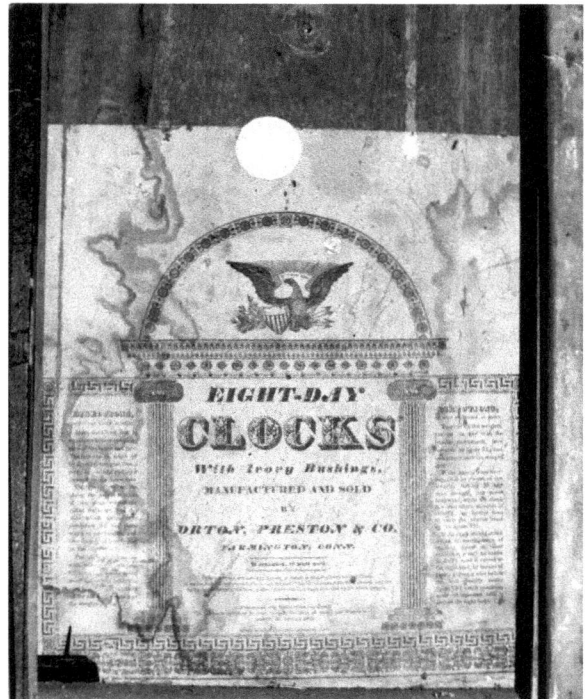

Figure 28. Slightly later "Eight-day" label of an Orton, Preston & Co. clock (not shown), but helping to interpret the badly torn label of Figure 27B. Note the printer's line is, "JOSEPH HURLBUT, PRINTER - HARTFORD, CONN". (Photo by Ward Francillon.)

Appendix A

Movement Identification Number Cross-Referencing

Movement	New No.	Old No.	Reference
30-hr. Wood	1.181	1.171	19
	7.115	5.112	19
	7.116	5.112	19
	7.153	5.112	19
8-day Wood	4.212	4.212	20
	4.213	4.213	20
8-day Strap Brass	4.112	4.112	22
	4.12	4.12	22
	5.112	5.12	22

Notes & References

1. Kenneth D. Roberts and Snowden Taylor. *Eli Terry and the Connecticut shelf clock.* 2nd ed. Fitzwilliam, NH: Ken Roberts Publishing Co., 1994.

2. *Seth Cowles vs. George Cowles et al.* Hartford County County Court Files, Box 423, RG3, Connecticut State Library; herein abbreviated as "the Seth Cowles case."

3. Farmington Land Records (FLR), Volume (V.) 43, p. 275, November 12, 1827.

4. Julius Gay (compiler). "Maps of Old Farmington." (unpublished bound volume of photostatic copies), Connecticut State Library. Figure 1 shows the Farmington village portion of the hand-drawn map designated "#9" in this collection. The configuration of the Farmington River, Great Brook, and "Pope Brook" (referred to as "Poke Brook" in the drawing, and in the "Historical Sketch of Farmington" by Richard Bissell Jr. (unpublished, 1927), are locatable in their respective positions on the U.S. Geological Survey's 7.5 minute topographic quadrangle map of New Britain, CT (1985), although the scale of the hand-drawn map is probably inexact.

5. Baker and Tilden. *Atlas of Hartford and Tolland Counties* [Connecticut]. 1869.

6. Lydia Hewes. *A short history of Farmington, Connecticut.* Farmington, CT: Committee of Connecticut's Tercentenary, 1964. According to this source, sometime prior to 1964, both the meat market and a red brick building located in front of it, formerly occupied by Colonial Pharmacy, were demolished to allow for the widening of State Route 4.

7. Christopher Bickford. *Farmington in Connecticut.* Canaan, NH: Phoenix Publishing, 1983. See also: Richard Bissell Jr. "Historical sketch of Farmington." (Unpublished pamphlet, 1927.)

8. Calvin Cowles (compiler). *The Cowles families in America.* New Haven, CT: Tuttle, Morehouse & Taylor, 1929.

9. Judson Deming (ed.). *Genealogy of the descendants of John Deming, Wethersfield, Connecticut.* Dubuque, IA: Mathis-Mets Co., 1904: 84 and 155.

10. Latter Day Saints, Family Search (online source). See also American Clock & Watch Museum research files.

11. Chauncey Jerome. *History of the American clock business for the past sixty years and life of Chauncey Jerome, written by himself.* New Haven, CT: F. C. Dayton Jr., 1860, reprinted by the American Clock & Watch Museum, 2001: 29. See also Arthur L. Brandegee and Eddy N. Smith. *Farmington, the village of beautiful homes.* (Hartford, CT, 1906); and Chris H. Bailey. *From rags to riches to rags: the story of Chauncey Jerome. NAWCC Bulletin* Supplement No. 15 (Spring 1986): 5.

12. Paul V. Heffner. "Ohio clockmaker: George Marsh." *NAWCC Bulletin*, No. 236 (June 1985): 270–279. See also: Dwight W. Marsh. *Marsh genealogy*. (Amherst, MA: Carpenter & Moorehouse, 1895.)

13. Chris H. Bailey. "Connecticut's clockmaking philanthropist, William L. Gilbert." In *Illustrated catalogue, William L. Gilbert Clock Co.* (August 1881; reprinted by the American Clock & Watch Museum, 1980): 174–189.

14. *William G. Bates vs. George Marsh & Co.*, Hartford County County Court Files, Box 444, RG3, Connecticut State Library.

15. *George Marsh & Co. vs. Harlow Wilcox*. Hartford County Superior Court Files, Box 83, RG3, Connecticut State Library.

16. Noah Preston Probate File, Madison County (IL) Probate Records, Illinois Regional Archive Repository (IRAD), Carbondale, IL.

17. Snowden Taylor. "Williams, Orton, Prestons & Co." Research Activities & News. *NAWCC Bulletin*, No. 264 (February 1990): 76–78.

18. See Appendix A, "Movement Identification Numbers."

19. Snowden Taylor. "Characteristics of standard Terry-type 30-hour wooden movements as a guide to identification of movement makers." *NAWCC Bulletin*, No. 208, Part I (October 1980).

20. Bryan Rogers and Snowden Taylor. "Eight day wood movement shelf clocks—their cases, their movements, their makers." *NAWCC Bulletin* Supplement No. 19 (Spring 1993).

21. Kenneth D. Roberts. *The contributions of Joseph Ives to Connecticut clock technology 1810-1862*. 2nd ed. Fitzwilliam, NH: Ken Roberts Publishing Co., 1988.

22. Snowden Taylor. "Connecticut style 8-day brass strap movements." Unpublished manuscript.

23. This clock was previously pictured in "Research Activities and News." *NAWCC Bulletin*, No. 328 (October 2000): 673.

24. D. R. Slaght. "Printers of Hartford [CT] 1825 thru 1860." Unpublished manuscript, NAWCC Library.

25. Paul V. Heffner et al. "Printers directory." Unpublished manuscript, a work-in-progress, NAWCC Research Committee.

26. Clarence Williams (compiler). "Washington Williams of Rocky Hill, Conn., together with his wife Mehitable Robbins, their forebears & descendants." Unpublished manuscript 1943–1944: Connecticut State Library.

27. Francis Atwater. *History of the town of Plymouth, Connecticut*. Meriden, CT: Journal Publishing Co.,1895: 353 and 355.

28. *Orton, Preston & Co. vs. Lucius Parsons et al.* Hartford County County Court Files, Box 427, RG 3, Connecticut State Library.

29. *Lemuel Whitman vs. Orton, Preston & Co.* Hartford County County Court Files, Box 443, RG3, Connecticut State Library.

30. J. Hammond Trumbull *Memorial history of Hartford County, Connecticut, 1633-1886*. Boston, MA. Edward L. Osgood, 1886: 126.

31. Carlos Bates papers, Box 2, RG 69:14, Connecticut State Library.

32. *Griswold & Pinneys vs. Orton, Preston & Co.* Hartford County County Court Files, Box 429, RG3, Connecticut State Library.

33. *George Peck vs. Orton, Preston & Co.* Hartford County County Court Executions, Box 533, RG3, Connecticut State Library.

34. Charles Henry Preston. *Descendants of Roger Preston of Ipswich and Salem Village*. Salem, MA: The Essex Institute, 1931. According to this source, John Stiles Preston, brother of Noah et al., "removed to Farmington Conn., and thence to Unionville, Conn., where he worked in the clock business". Additional circumstantial evidence of John S. Preston's involvement in Williams, Orton, Prestons & Co. will be presented in Chapter 4.

35. Eli Terry Jr. merchant letterbooks, American Clock & Watch Museum.

36. Chris H. Bailey. Letter dated September 6, 2006, "Williams, Orton, Prestons & Co." American Clock & Watch Museum research files.

37. *Hartford Courant*, August 24, 1909.

38. Vital Records Index, Connecticut State Library.

39. Farmington Land Records (FLR), V. 47, pp. 352–353, May 6, 1842.

Chapter 2

Unionville Firm: Seymour, Williams & Porter

By the early 1830s, it was becoming clear that Farmington village, lying in the vicinity of the Farmington River's floodplain, was better suited to agriculture than manufacturing.[1] Meanwhile, an entirely different scenario was taking shape in the northwestern portion of town, in the tiny settlement soon to become known as Unionville. So called by virtue of its origin as the Union School District, the village was located at the junction of the towns of Avon, Burlington, and Farmington. According to local historian Mabel S. Hurlburt:

> Except in Union District, no mention is made of Unionville in any of the old record books prior to 1834. Then the United States Government established a post-office here and this locality received its name. Long prior to this date we hear frequent allusions to Langdon's

Quarter and come across the name of one of the earliest settlers - Solomon Langdon.[2]

By the late 1820s, Solomon Langdon (1749–1835), farmer, innkeeper, and owner of extensive property and several mills within the sparsely populated area,[3] began selling off portions of his holdings. One eager early buyer was a man named Edward Seymour, who was then in his late twenties.

Seymour, an experienced clock casemaker, had spent his early career in the large, successful clock factories of Plymouth, the birthplace of the wooden shelf clock movement. He was born in Farmington on March 27, 1801, to parents Luther and Rebecca (Curtiss) Seymour.[4] It is likely that he received training as a joiner from his father, a skilled cabinetmaker and housewright of substantial local repute.[5] Prior to establishing his own clock

Figure 29. Former home of Edward Seymour, on Main St., Unionville, from a postcard ca. 1900. (Courtesy American Clock & Watch Museum.)

factory, Edward Seymour had made cases for clockmakers Seth Thomas and Eli Terry,[6] the latter being the great innovator who had pioneered the wooden shelf clock movement, and who held the patents thereon. Now, with the clock industry booming in west-central Connecticut, Seymour was ready to strike out on his own. In April 1834, while Seymour completed work on his new home on what would later become Unionville's Main Street, he was appointed the village's first postmaster, a position he held for the next five years. The post office occupied a part of Seymour's home[7] (Figure 29).

The need for a post office in the village may well have arisen from the complex business Edward Seymour had entered into, which required communication with far-flung regions of the young American nation. It was also a sign of the village's rapid growth. Despite its expansion, however, Unionville remained part of the town of Farmington until 1921, when it was incorporated as a borough. Consequently, even as the village developed quickly during the 1830s and 1840s, the designation "Unionville" appeared only occasionally in the local records.[8]

Seymour, Williams & Porter

The great advantage of the largely uninhabited northwest corner of Farmington was that both the Farmington River and its tributary, Roaring Brook, flowed through the area. By 1829, a feeder dam and canal had been completed there, for the purpose of diverting water from the river to augment flow in the newly built Farmington Canal at the center of Farmington village. This arrangement allowed canal boats to travel upstream as far as the wharf and store that the brothers James & Augustus Cowles had constructed in what would become Unionville center.[9] In 1830, Thomas and Joshua Youngs of Farmington, John T. Norton of Albany (at the time), and Abner Bidwell of New York City, together with the Cowles brothers, incorporated as the Farmington River Manufacturing Co. They constructed a second dam on the river, two miles upstream of the feeder dam, further unlocking Unionville's potential (in those days before the widespread availability of steam and rail) for power, transportation, and growth. Subsequently, the firm offered leases for water power sites suitable for manufacturers. All this helps explain how Edward Seymour was able to obtain financial backing during—and even beyond—the boom years that preceded the great nationwide financial Panic of 1837.

Indeed, Edward Seymour, the first named partner in the clock-making firm of Seymour, Williams & Porter, was not only one of the earliest to envision the benefits of Unionville for manufacturing, but also one of the earliest to follow the lead of Farmington's wealthy merchants by investing in land in the village, borrowing heavily to do so. On November 26, 1829, about one month prior to commencing his soon-to-be-mentioned land purchases, Edward Seymour—then technically a resident "of Torrington"—married Harriet Johnson "of Bristol" at Litchfield.[10] The couple moved to Unionville in 1830 or 1831. Seymour spent the remainder of his life as a clockmaker, businessman, peddler, traveler, pioneer, and adventurer.

By means of two deeds, both recorded on December 30, 1829,[11] Solomon Langdon transferred two parcels located in Farmington to Edward Seymour, then of Plymouth, in Litchfield County. In the first, for $123.12, Langdon:

> ...s[old]...unto...Edward Seymour a piece of land in said Farmington, for a mill seat or water privilege...[the parcel's meets and bounds are described by locations of "black oak staddle", "stones", "chesnut sprouts" [sic], "hemlock staddle", "chesnut staddle", "white oak staddle", "apple tree", "white oak", "maple", and "yellow oak staddle", on both sides of the brook]...containing six acres & twenty five rods - be the same more or less: - & bounded every way on my own land: with the right of conveying the water from his mill or mills, hereafter to be built, over and across my land to the brook [i.e., Roaring Brook], in the most convenient place..:[12]

Including some reservations and restrictions, the deed conveys a sense of the area's then undeveloped character. Reference 1 (p. 300) says that Seymour purchased Solomon Langdon's interest in a gristmill here, but the deed does not say that. In the second deed Langdon conveyed to Seymour what was likely a one-acre home site, "near the school house in Union District...bounded Southward, on highway, and in every other direction, on my own land."[13]

Seymour's next target was Langdon's woolen mill. Langdon had previously leased the site to Eli Wood,[14] and various conditions and restrictions placed on it had been passed on to the two separate current half-owners, John Stanley and Abraham Clark. Each of these parties sold his separate half mill shares to Seymour for $267.50, and $267.00, respectively, on August 14, 1830, the deeds stating, (following Stanley's):

> ...one equal undivided half of the fulling mill with a like proportion of all the buildings, tools, implements, and apparatus attached to or in any way belonging to the clothier's works; also the same proportion of all the carding machinery and the buildings covering the same.[15]

Each of the deeds also granted Seymour one half of the rights and privileges formerly granted to Eli Wood. On November 27, 1830, in return for $33.34, Langdon sold Seymour "the land on which the Fulling Mill, and the other buildings connected therewith now stand, and which have been used for Clothier's works, and for carding wool" and also:

> ...the right and privilege of building a Dam across the brook [Roaring Brook] running through my land, and of bringing the water from said brook on my land, in the channel, where the same has hitherto been brought, to the said mill and other buildings...[together with the] right and privilege of using said mill, and buildings...for such purposes as he...may think proper.[16]

Not only was Edward Seymour accumulating and developing property in the center of the village that would soon become known as Unionville, he was also becoming involved in a complex business venture, documented in a two-volume set of account books that belonged to the firm of Seymour, Williams & Porter, now in the collection of the Connecticut Historical Society. In the first volume of the accounts, one left-right pair of pages is headed "Dr [i.e., Debit] Seymour Williams & Porter in account with A. F. Williams Treasurer Cr [i.e., Credit]". The earliest entry in this volume is dated February 16, 1831. The "first part" of the accounts is set up in normal daybook form, but many clock firms at this date were small, and did not typically have separate accounts maintained by treasurers! The second part of the accounts is set up as a series of lists of notes due to and from the firm, with the earliest item noted therein dated February 12, 1831. Hence, the firm of Seymour, Williams & Porter must have dated to February 1831, or earlier.[17]

Austin F. Williams (1805–85),[18] the second named partner in the firm of Seymour, Williams & Porter, married Timothy Cowles's daughter Jeanette (1810–70), on September 11, 1828.[19] Thus Williams, whose own business interests (as we shall see) extended well beyond Unionville village, to Farmington village, Plainville village, and New York City, served as Cowles's financial agent (and personal representative) in matters relating to the clock making firm.

Timothy Porter (1804–76),[20] the third named partner in Seymour, Williams & Porter, also a joiner, was the son of Selah and Susannah (Cadwell) Porter of Farmington.[21] Timothy Porter's father was the "Captain Selah Porter," for whom clockmaker Chauncey Jerome had worked to help build a new house in Farmington village for Major Timothy Cowles in 1815, which Jerome described as "then the best one in Farmington."[22] On December 26, 1833, Timothy Porter married Emeline Phelps,[23] daughter of innkeepers Noah L. and Jemima (Stedman) Phelps, also of Farmington.[24]

During the summer of 1831, builders Ambrose Hart and Chauncey Moses of Farmington and Joshua Moses of Burlington brought suit against Edward Seymour and Timothy Porter in Hartford County County Court, accusing the two of failing to completely pay for timber, framing, stonework, and for hewing a door sill for a new building, likely a clock factory west of Roaring Brook. The builders' bill, dated February 12, 1831, totaled $361.66. An entry in the firm's accounts dated February 16, 1831, debited Seymour, Williams & Porter in the amount of $300 "to cash Rec'd from T. Cowles per note". This entry was offset by a credit in the same amount "by cash Paid T. Porter" dated the same day. According to the court records, Seymour and Porter paid $300 of the bill from Hart, Moses, and Moses on February 17, evidently with the money obtained from Timothy Cowles. Seymour and Porter made another $13.61 in payments, including a payment of $10.00 via Timothy Cowles, on March 3.[25] The court rendered judgment in favor of plaintiffs Hart, Moses, and Moses, in the amount of $23.09 damages, during its August 1831 term.[26]

In the spring of 1831, Solomon Langdon sold Edward Seymour several more small parcels of land, the first deed, dated March 26, 1831, mentioning "the new factory."[27] On the same day, Seymour sold one small parcel back to Langdon.[28]

On June 11, 1832, Edward Seymour transferred to his partners Austin F. Williams and Timothy Porter, each separately, a one-third part of all the property deeded to him by Langdon and others as previously described.[29] Williams and Porter each paid $239.15 for his share. The brook in the deed was now designated "roaring brook," but there was no mention of "Union" in any form. These deeds were also described in Reference 1 (p. 300), but with some errors, and the quote from the deeds should be corrected to read "...remise, release and forever quit-claim...all such right & title as I...have, in or to one equal undivided third part of the several pieces of land hereinafter described, with the buildings thereon..." On November 9, 1832, Langdon sold Seymour, Williams, and Porter together, for $10.00, 20 rods of land, "bounded south on the highway, and East north and west on my own land - a brook running through the same."[30] This was a small but important piece of land in terms of access, where Roaring Brook met a new section of road.

Although added onto several times during the intervening years, a building on the east bank of Roaring Brook, and northwest of the intersection of Farmington Avenue and West Avon Road in Unionville, has been identified as Seymour, Williams & Porter's easternmost clock factory building.[31] According to the same source, by 1855 the building was occupied by Edward K. Hamilton's Hook & Eye Manufactory (Figure 30). Figure 31 shows a photograph of the building taken circa 1885, when it was occupied by a building materials firm, Sanford & Hawley.

On the basis of the information provided in the deeds and in the firm's extant accounts, Seymour, Williams & Porter probably began producing clocks during 1831, or at the latest, 1832. The first were short-pendulum varieties,[32] with labels reading (in part), "PATENT CLOCKS / INVENTED BY / ELI TERRY, / AND / MANUFACTURED / BY / SEYMOUR, WILLIAMS & PORTER, / FARMINGTON, CONN." See Figures 32A–32C. These have 30-hr. wooden movements, apparently made by one of Eli Terry's firms, presumably E. Terry & Sons. Next came long half-pillar and splat clocks, with Seymour, Williams & Porter's own movements with 30-tooth escape wheels and escape wheel bridges cut off square at the tips. An example is shown in Figures 33A–33D. Its movement (Type 1.151[33]; Figure 33C), is set into the clock's backboard (Figure 33D), as in clocks of the Terry family firms, seeming to reinforce the evidence that Edward Seymour had once worked for Eli Terry. Then followed other long clocks with Seymour, Williams, & Porter's own Type

1.411 movements,[34] with escape wheel bridges rounded (but not fully round) at the tips. See Figures 34A–34C. Some of Seymour, Williams & Porter's 30-hr. clocks had unusual external wooden alarm mechanisms, as seen in Figures 35A–35C. There were also 8-day wooden movement clocks, based on the E. Terry & Sons plan, as in Figures 36A–36C. Note the seconds indicator on the upper portion of the dial in Figure 36A. Some of the firm's 8-day clocks also had external alarms. An example of the latter is shown in Figure 37.

A clock casemaker and carver, R.[odolphus] E. Northrup of New Haven, CT, who produced a number of shelf clocks with both wood and brass movements under his own name, sourced some 30-hr. wood movements from Seymour, Williams & Porter. At least one example that bears Northrup's own label inside is known, but Seymour, Williams & Porter's label is found on the back of the clock case (Figures 38A–38D), an apparent effort to credit the latter firm as the clock's movement maker.

A few shelf clocks in a variety of well-constructed cases, mostly with wood movements by various makers, are also known, with labels stating: "CLOCKS, / MADE BY / SEYMOUR, WILLIAMS & PORTER. / CASED AND SOLD BY / CARTER & WELLER, / STOCKBRIDGE, MADISON COUNTY, N.Y." One example is shown in Figures 39A and 39B. It is strange that the short-lived firm of Carter & Weller is known to have been in business only from 1839 to 1841, by which time (as we shall see) Seymour, Williams & Porter had been dissolved for about four years. Stranger still is the fact that none of Carter & Weller's clocks have yet been observed to contain a movement made by Seymour, Williams & Porter![35] The movement shown in Figure 39B, for example, was made by the firm of E. & G. W. Bartholomew, or G. W. Bartholomew, of Bristol, CT. The label from a similar example is shown in Figure 40. Little is known about Jonathan Carter (1813–80) and Daniel S. T. Weller (ca. 1812–?), the partners in Carter & Weller, who both left Stockbridge by 1842, with their finances in ruin,[36] and nothing is known about the nature of their relationship with Seymour, Williams & Porter.

As of 1832, only about 20 families resided in Unionville.[37] During the following year (1833), Edward Seymour served as one of five members of the "Committee of Arrangements," organizers of the village's Independence Day celebration. According to a newspaper account of the event, after a reading of the Declaration of Independence at 2 p.m., about 100

Figure 30. Portion of the village of Unionville, showing "Hamilton's Hook & Eye Manu[factor]y," formerly Seymour, Williams & Porter's eastern clock factory building, east of Roaring Brook and just west of the road leading north toward Avon. (Source: E.M. Woodford, Smith's Map of Hartford County, Connecticut 1855.)

Figure 31. Photograph of Sanford & Hawley building (former eastern clock factory building of Seymour, Williams & Porter), ca. 1885. (Courtesy American Clock & Watch Museum.)

Figure 32A. Seymour, Williams & Porter short pendulum shelf clock, exterior view.

Figure 32B. Label of Seymour, Williams & Porter short pendulum shelf clock.

Figure 32C. Movement of clock shown in Figures 32A and 32B, thought to be made by Eli Terry & Sons. (Figures 32A–32C, photos by Ward Francillon.)

Figure 33A. Seymour, Williams & Porter long pendulum, half-column and splat-style shelf clock, exterior view.

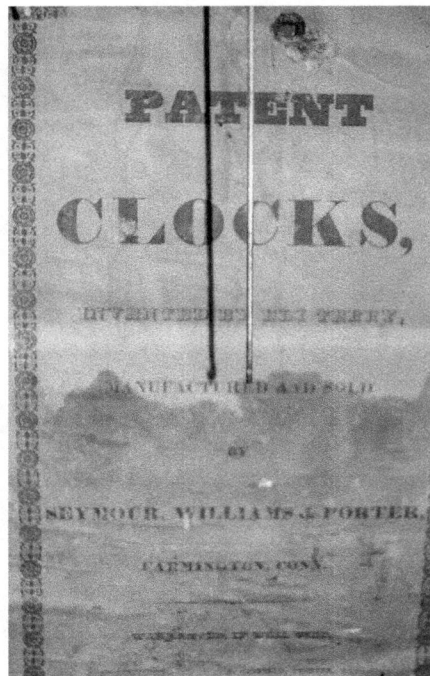

Figure 33B. Label of clock shown in Figure 33A.

Figure 33C. Movement of clock shown in Figures 33A and 33B, made by Seymour, Williams & Porter. Note 30-tooth escape wheel and escape wheel bridge cut off square at tip.

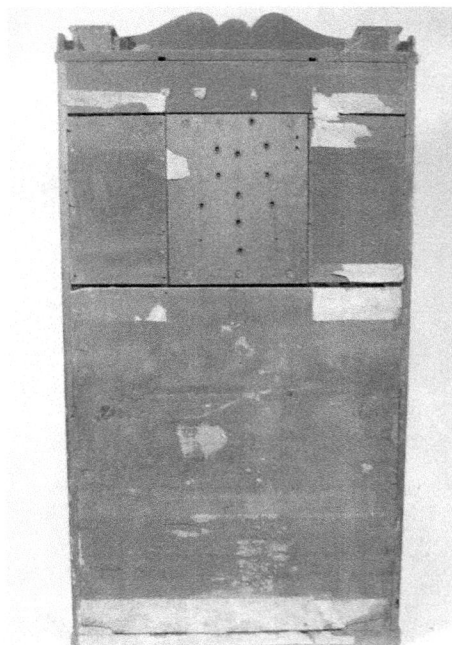

Figure 33D. Back of clock shown in Figures 33A–33C. The movement is set into the backboard, a feature typical of early Terry firms. (Unless otherwise noted, all photos by MJD, courtesy authors' collections.)

Figure 34A. Seymour, Williams & Porter long pendulum clock, exterior view.

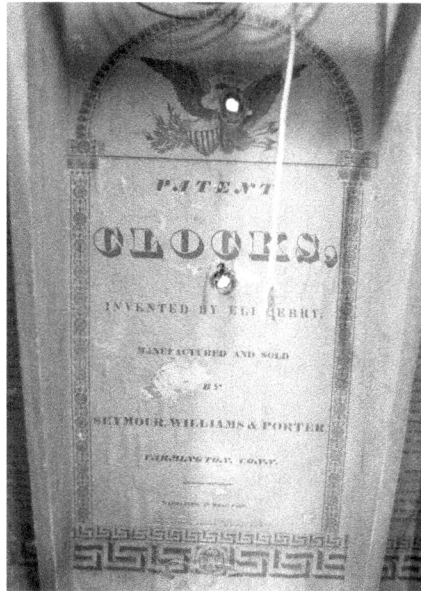

Figure 34B. Label of clock in Figure 34A.

Figure 34C. Movement by Seymour, Williams & Porter, of clock shown in Figures 34A and 34B. Note 30-tooth escape wheel and escape wheel bridge more rounded at the tip than the example shown in Figure 33C, but not fully round.

Figure 35A. Seymour, Williams & Porter shelf clock, with 30-hr. time, strike, and outboard alarm wood movement, exterior view.

Figure 35B. Label and outboard alarm movement of clock shown in Figure 35A.

Figure 35C. 30-hr. wood movement by Seymour, Williams & Porter, of clock shown in Figures 35A and 35B. (Figures 35A–35C courtesy of Chris Brown; Gene Gissin photos.)

villagers "sat down to dinner under a bower."[38] When the cloths covering the tables were removed, 22 toasts, mostly of a patriotic nature, were drunk. It is revealing to note that, when Unionville's exuberant inhabitants raised their glasses for the 13th time that day, the villagers drank to: "Water power, steam power, horse power, dog power, any power that will drive machinery..."[39]

On September 25, 1833, Edward Seymour purchased property from Elijah O. Gridley of Farmington, described as follows:

> ...one certain Sawmill...with the tools, implements and apparatus thereto belonging, together with the yard belonging to said mill, described as follows viz: Beginning at the south line of the Highway 64 links west of the front of the Westerly Abutment of the bridge which crosses the Mill pond belonging to said Mill running south...to the [Farmington] river - then following the river easterly to the mouth of the [Roaring] brook - then following the brook, to the North end of the Pentstock [sic], then by the west bank of the pond to the highway - thence by the highway to the first mentioned boundary: - Bounded North by the highway and the Mill pond - East by the middle of the brook, as it now runs - South by the Farmington river and Solomon Langdon: West, by James and Augustus Cowles...[40]

The sale price was $350. This location was north of the Farmington River, on the west side of Roaring Brook, southwest of the road to Farmington and the mill pond near the mouth of the brook. Seymour divided the sawmill property with his partners in two deeds, on May 2, 1834, each paying $117. See Reference 1 (p. 300), but the wording of the deeds should be corrected to read as follows:

> ...one equal undivided third part of a certain Saw-Mill standing near the dwelling house of Solomon Langdon...with the tools, implements, and apparatus thereto belonging: together with the yard belonging to said mill, as described in a deed from Elijah O. Gridley to me the said Seymour dated September the 25th AD 1833...[41]

By the spring of 1829, such was the unprecedented speed of travel on the Farmington Canal that Elijah O.

Gridley's schedule as the pilot of the canal boat *DeWitt Clinton* included: "...[leaving] Farmington on Thursday of each week, returning, leaving New Haven (i.e., the terminus of the Canal, at a distance of 41 miles from Farmington), on each succeeding Thursday..."[42] Gridley will appear again later, as a lender to several other Unionville clock makers.

Records relating to a number of court cases offered further insights into the activities of Seymour, Williams & Porter, and, as we will see, into those of Edward Seymour's subsequent clock-making firm and its successors. For present purposes, seven of the court cases, with references, are summarized in Table 2A, each involving the firm of Seymour, Williams & Porter as either plaintiff or defendant (Group I; Cases A–G). The remainder of the court cases, which involve the firm's successors, will be taken up in the next chapter. The authors will refer to Table 2A often when commenting on the cases.

Taken as a whole, the court cases in Group I, Table 2A, hint at Seymour, Williams & Porter's declining fortunes. In the first four cases listed, the firm was the plaintiff; in the rest of the cases it was the defendant. In Cases B and C, defendants William Huntington and Rufus King were taken to gaol [jail]. Samuel Deming, the plaintiff in Case G, appeared previously, as a partner in the firm of Cowles, Deming & Camp, whose mill in Farmington village had been rented by George Marsh's clock-making firms, and likely by their successor firm, Orton, Preston & Co., prior to 1835. Here he was a provider of unidentified goods or services to Seymour, Williams & Porter. William Bradley (1806–72) was Edward Seymour's "agent" in Case E, which actually consists of a combination of three lawsuits initiated by Timothy Cowles on the same day. William Bradley's wife, Adaline (Johnson), was the sister of Edward Seymour's wife.[43] More on Bradley later.

Extant records of Court Case D, Table 2A, *Seymour, Williams & Porter vs. Thomas I. Lewis, Charles Cowles, & Samuel Brown* provide interesting details. On April 22, 1836, the constable levied execution of the Court's judgment in favor of Seymour, Williams & Porter, on "...17 boxes of Clocks, with the boxes of weights with the said boxes marked Lewis & Brown, Portage, Ohio, the property of the within debtors..." The boxes were sold "at the end of twenty days at public vendue [sic]." On that date, the auction having been announced by the customary drumbeat, the results of bidding were as follows:

Table 2A. Court cases involving the firm of Seymour, Williams & Porter (SWP).

The date of the writ and summons initiating the court case(s) is provided under "Date of Writ(s) (or Execution(s))". If, instead of a writ, the only available record of the court case is related to the execution of a court's judgment in the case, the date(s) on which the court granted the execution(s) are provided in parentheses. If the dispute was over non-payment or partial non-payment of a note, the amount of the note is given under "Amt. of Note"; the date of the note is listed in parentheses below. "Outcome / Additional Info" is the outcome of the case as far as known, followed by a brief mention of any unusual aspect of the case in parentheses. The full reference for each court case is provided under "Reference". "Exn." = execution; "plff" refers to "plaintiff"; "defts" = defendants; "gaol" = a British term for "jail"; "CSL" = Connecticut State Library; "CHS" = Connecticut Historical Society; "Htfd. Co." = Hartford County; "Unk." = Unknown. All amounts are in dollars.

Case Date of Writ(s) (or Execution(s))	Title of Case	Amt. of Note / (Date of Note)	Outcome / (Additional Info)	Reference
		Group I. Seymour, Williams & Porter (SWP)		
A (11/14/1833)	*SWP v. James Farnsworth*	[Unk.]	Exn. in favor SWP.	Austin F. Williams papers, CHS.
B (11/19/1833 12/9/1837)	*SWP v. William Huntington* (continuation of same case)	[Unk.]	Agreed to pay in 3 wks.; on 2nd Exn. Deft. to gaol.	Htfd. Co. Co. Court Exns., Box 530, RG3, CSL.
C (11/13/1834)	*SWP v. Rufus King & Samuel Allen*	[Unk.]	Exn. unsatisfied; R. King to gaol.	Htfd. Co. Co. Court Exns., Box 530, RG3, CSL.
D (3/30/1836; 9/18/1839)	*SWP v. Thomas Lewis, Charles Cowles & Samuel Brown*	[Unk.]	Exn. in favor SWP unsatisfied; 2nd Exn. (See text.)	Htfd. Co. Co. Court Exns., Box 530, RG3, CSL.
E 10/7/1837 (1/31/1834)	*Timothy Cowles v. SWP*	31,000.00	Defts. defaulted; Exn granted on 4/23/1838; plff awarded $16,903.30; 2nd Exn. 6/30/1838.	Htfd. Co. Co. Court Files, Box 434, RG3, CSL.
F 7/29/1839	*Timothy Cowles v. SWP* (X3)	1860.00 (1/1/1835) 1701.60 (1/1/1836) 550.00 (7/1/1836)	SWP paid part of debt note by note; Exn. 12/6/1839 (X3) (Wm. Bradley agent; E. Seymour out of state but may request new trial; 3rd note signed by ES & AFW.)	Htfd. Co. Co. Court Files, Box 439, RG3, CSL.
G 10/28/1841	*Samuel Deming v. SWP*	300.00 (2/25/1836)	Exn. in favor plff on 11/11/1841.	Htfd. Co. Co. Court Files, Box 443, RG3, CSL.

2 boxes, or 12 clocks, to Gad Cowles [at] 3.90	=	46.80
9 boxes, or 54 clocks, to Timo[thy] Porter at 3.50	=	189.00
6 boxes, or 36 clocks, to Gad Cowles at 3.97 ½	=	<u>189.90</u>
17 boxes, or 102 clocks, - amounting to		$425.70[44]

The sum of money raised by means of the auction was put in part payment against the total court award of $3,105.61 plus $13.86 in court costs in favor of the plaintiff, Seymour, Williams & Porter. The firm of Lewis & Brown appeared frequently in the Seymour, Williams & Porter accounts, although it is unclear whether the clocks taken in the execution had actually been made by the firm. Furthermore, the court case demonstrates that Timothy Porter, as a purchaser of nine of the boxes of clocks, was still involved in the clock trade as late as April 1836. In view of the events outlined below, this is somewhat surprising.

Leaving our discussion of the court cases in Table 2A for a moment, Reference 1 (p. 302) reported an important deed from Timothy Porter to his two partners in Seymour, Williams & Porter, namely, Edward Seymour and Austin F. Williams, but made errors in the date and the deed wordings, and omitted the price. The actual date was January 29, 1835, the sale price was $100 jointly, and the correct wording was:

> ...all...lands, with the buildings thereon, conveyed to me by three deeds herein after described [i.e., FLR Vol. 44, p. 547; Vol. 45, p. 81; and Vol. 45, p. 139] including the tools, implements, and apparatus belonging to the saw-mill, but exclusive of the machinery and tools belonging to the other buildings...all of which land and buildings are situated in said Farmington, at a place called Unionville.[45]

In short, Porter abruptly sold his entire land holdings in Seymour, Williams & Porter to his partners. A notice that the firm of Seymour, Williams & Porter was dissolved effective February 2, 1835, appeared a few days later, on page 3 of the *Connecticut Courant* (Figure 41). Immediately following this was an additional

"NOTICE", signed "A.F. WILLIAMS", stating that: "The subscriber, having formed a connection in the Dry Goods Business [in] New-York..." and gave directions for settling accounts with Williams "at the store." Oddly, the second notice contained no mention of the clock business.

So, although Edward Seymour was to continue the clock business "at the same Establishment," the firm of Seymour, Williams & Porter was no more. But in the early nineteenth century it took a long time to unwind a business. Examination of the "first part" of Seymour, Williams & Porter's accounts, mentioned previously, shows that the firm was fully active during the years beginning in 1832 through 1834, but less so after that time, consistent with the firm's dissolution on February 2, 1835.

Thereafter, Timothy Porter continued to pursue his career as a joiner. On August 10, 1835, the *Connecticut Courant* carried Porter's advertisement for four journeymen joiners. The ad went on to say: "The subscriber has on hand a lot of CLOCKS, of *Seymour, Williams & Porter's* manufacture...", which Porter offered for sale "low for cash or approved credit" (Figure 42). On March 14, 1836, Porter again advertised in the *Courant*, seeking "...several *JOURNEYMEN JOINERS*, to whom good wages will be given..." and giving a Farmington location. Although he continued to deal in clocks occasionally, the authors found no evidence that Timothy Porter (if he had ever been more than an investor in the firm) was involved in producing clocks or clock cases after the dissolution of Seymour, Williams & Porter.[46] He remained a Farmington resident. However, January 10, 1837, found Porter in Macon, GA, installing an elegant interior for a new church there, ordering the pulpit, bannisters, and trim from Leonard Winship's cabinet shop in Farmington, all based on the design of the pulpit and interior of Farmington's Congregational Church.[47] As we will see, it is possible that Timothy Porter's involvement in the construction of several additional public and private buildings in Macon was connected with a clock sales venture.

A closer look at Seymour, Williams & Porter's account books is warranted. The daybook-like "first part" of the accounts, "Seymour, Williams & Porter in account with A.F. Williams treasurer," actually mentions both a daybook and a ledger, so in addition to the two extant volumes, more familiar types of accounts were kept. Most of the available entries are abbreviated, usually providing a single name or single item, but some are slightly

Figure 36A. Seymour, Williams & Porter 8-day shelf clock, exterior view.

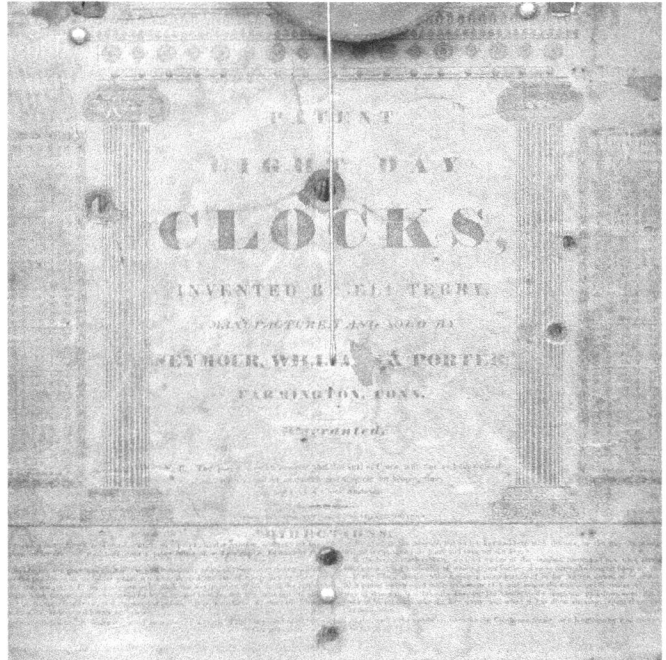

Figure 36B. Label of clock shown in Figure 36A.

Figure 36C. Seymour, Williams & Porter, Terry-style 8-day wood movement of clock shown in Figures 36A and 36B. (Figures 36A–36C courtesy of Chris Brown, photos by Gene Gissin.)

Figure 37. Another example of Seymour, Williams & Porter "Terry-style" 8-day clock, with door open, showing outboard alarm mechanism. (Courtesy private collection, Ward Francillon photo.)

Figure 38A. Exterior of carved case shelf clock by R. E. Northrop, New Haven, Conn.

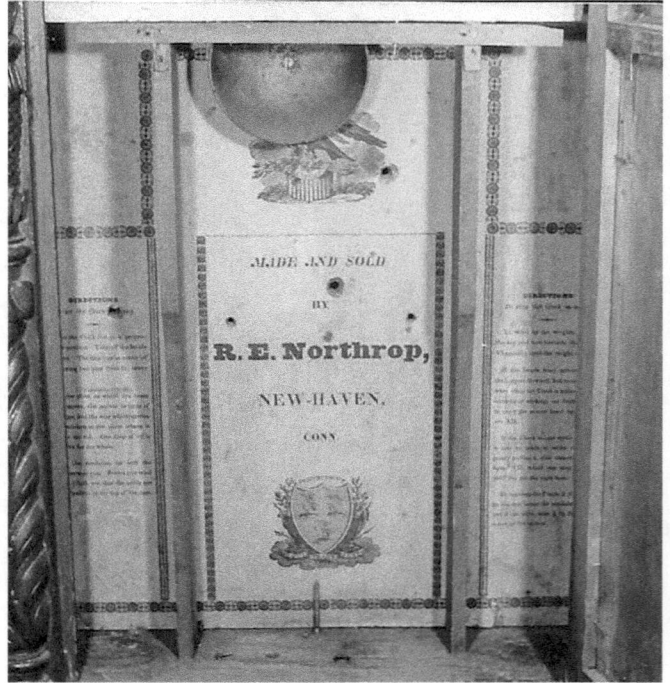

Figure 38B. R. E. Northrop label inside clock shown in Figure 38A.

Figure 38C. Seymour, Williams & Porter 30-hr. movement in clock shown in Figures 38A and 38B.

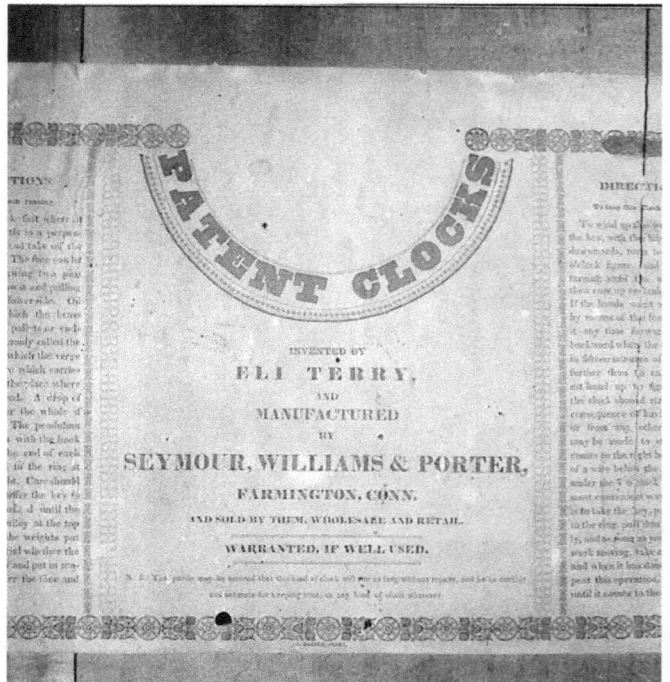

Figure 38D. Seymour, Williams & Porter label on backboard of R. E. Northrop clock shown in Figures 38A–38C. (Figures 38A–38D courtesy private collection, photos by Ward Francillon.)

Figure 39A. Exterior of clock with broken tablet. Its label reads: "...MADE BY / SEYMOUR, WILLIAMS & PORTER / CASED AND SOLD BY CARTER & WELLER, / STOCKBRIDGE, MADISON COUNTY, N.Y."

Figure 39B. Type 10.147 movement (Snowden Taylor, unpublished), made by E. & G. W. Bartholomew, or G. W. Bartholomew, of Bristol, CT, inside clock shown in Figure 39A. (Figures 39A and 39B courtesy of Chris Brown, Gene Gissin photos.)

more complete. In these accounts, incoming cash and notes seem to be entered as "Dr to" items, while outgoing ones are entered as "Cr by" items, as if Seymour, Williams & Porter was the subsidiary of a larger entity. The "second part" of the firm's accounts consists of a series of lists: (1) "Accounts and Notes payable & when due," (2) "Accounts (and Notes) Paid & when paid," and (3) "Notes and a/ct. due from Seymour, Williams & Porter," but some pages are for notes and accounts due, some are for notes and accounts paid, and some for notes given, all in a somewhat confused order and manner. Certain items from these second part accounts can be matched to the first part accounts, although many cannot. The tables below have been compiled mostly from the "first part" of the accounts, supplemented with information obtained from the "second part."

Table 3 lists clockmakers (plus one clock worker) recognized by the authors but undesignated as such in the accounts. Also shown are the amounts paid to or received from each. The few "received from" items are

given in parentheses. First, note Eli Terry, $100, which is marked in the accounts, "for patent." That is a payment to Terry for the right to make his patent wooden movements! Only two such payments to Terry were previously known: one from Ephraim Downs in 1830 for $469.60 (Reference 1, p. 224) and one from Hopkins & Alfred in 1833, in the form of 20 clocks (or $100, if clocks were $5) (Reference 1, p. 286). Elsewhere in Table 3, the 1832 date for "Marsh & Gilbert," presumably Marsh, Gilbert & Co., agrees with the authors' findings in Chapter 1. We do not know what services "Marsh & Gilbert," Ephraim Downs, or Atkins & Downs provided. Most clockmakers who received a single payment (single date) probably performed some single task for Seymour, Williams & Porter. Others were involved over relatively long periods of time; for example, John B. Terry (see Reference 1, p. 243), Julius Peck,[48] Lynde Preston, Bradley & Comstock,[49] and F.[rederick] W. Crum. It is likely that some of them worked onsite as clockmakers.

George Barber, F[rederick] W. Crum, and Asa B. Darrow, all encountered in the accounts and mentioned in Table 3, later produced small numbers of shelf clocks at Unionville, either alone, or in partnerships with others, and play prominent roles in later chapters. At this point, two notes given by the firm of Seymour, Williams & Porter found in extant papers of Austin F. Williams, now in the collection of the Connecticut Historical Society, are of interest, because they document payments to George Barber and Frederick W. Crum, respectively, "for value received," in 1835. The earlier of the two notes, for $165.30, was given to George Barber on March 23, 1835. According to the note, Barber was not paid the money he was owed until October 25, 1836. The second of the two notes, given to Frederick W. Crum on June 1, 1835, was simply marked "Paid." Considered together with the information provided in Table 3, therefore, Crum and Barber had each been either an employee of, or a supplier to, Seymour, Williams & Porter.

Returning to our discussion of Table 3, "Burton Paine" may be Joseph Burton Payne, a minor clock purveyor (Reference 1, p. 351). "L. Smith" was undoubtedly Levi Smith, a Bristol clockmaker during the 1840s (Reference 1, pp. 338, 340, and 351) but a Farmington resident and justice of the peace during the 1830s. Smith may also have been working in Seymour, Williams & Porter's clock factory.

"Lynde Preston" (no known relation to the brothers Noah and Eli D. Preston) was undoubtedly the Lynde B. Preston who became a clock factory employee of H. Welton & Co. of Plymouth during the 1840s.[50] In the two "received from" entries, Ransom Smith was paying on a note, and Comstock & Bradley of Plymouth (undoubtedly Lucius Bradley and LaFayette Comstock; the firm name Bradley & Comstock has been observed on labels of 30-hr. wood movement shelf clocks originating in Plymouth [Reference 1, p. 349]), could have been doing the same. Ransom Smith was involved in clock sales and perhaps manufacturing, first in Hartford,

Table 3. Cash or notes that Seymour, Williams & Porter either paid to or received from clockmakers. "Received from" items are in parentheses.

Name	Amount paid to (or from)	Date
Atkins & Downes	$344.50	March, 1832
Geo.[rge] Barber	75.00	September 1836
Bradley & Comstock (also Comstock & Bradley)	1100.51 (105.50)	June 1834–June 1835 June 1834
F. W. Crum	374.00	June 1834–August 1836
Asa B. Darrow	50.00 or 60.00	April 1835
Ephraim Downs	286.00	September 1831–May 1832
C. & L.C. Ives	30.00	June 1834
Messrs. Marsh & Gilbert	83.93	November 1832
Burton Paine	5.00	November 1832
Julius Peck	785.37	August 1832–July 1833
Lynde Preston	123.91	September 1833–December 1836
L. Smith	6.20	June 1833
Ransom Smith	(65.50)	July 1832
Eli Terry	100.00	January 1832
John B. Terry	450.96	October 1831–June 1835

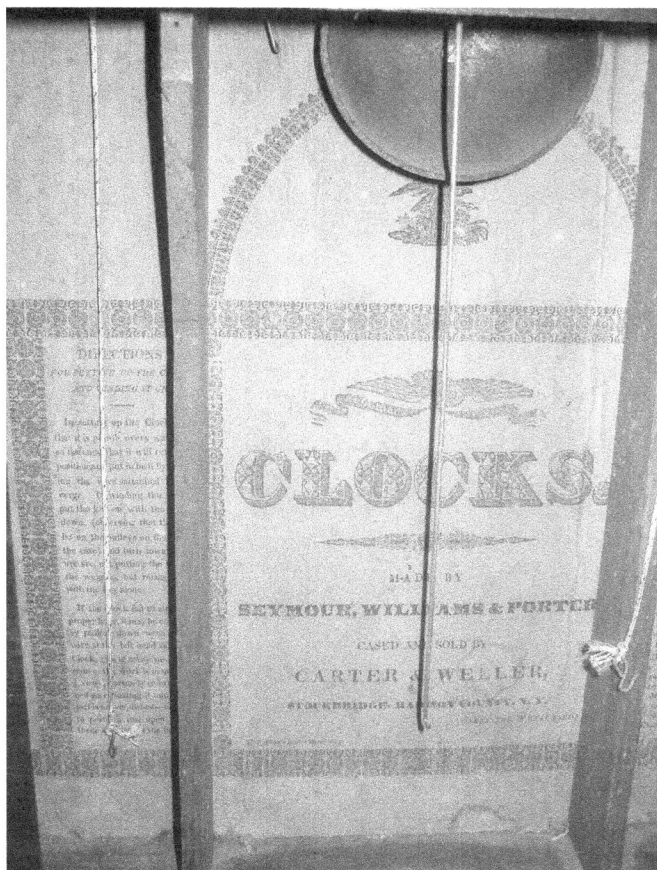

Figure 40. Label from another example of a shelf clock cased and sold by Carter & Weller, but crediting Seymour, Williams & Porter as its maker. (Courtesy of Russ Oechsle.)

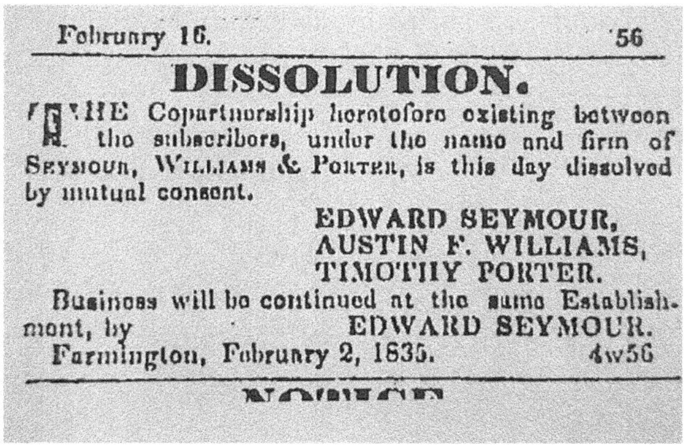

February 16. '56

DISSOLUTION.

THE Copartnership heretofore existing between the subscribers, under the name and firm of SEYMOUR, WILLIAMS & PORTER, is this day dissolved by mutual consent.

EDWARD SEYMOUR,
AUSTIN F. WILLIAMS,
TIMOTHY PORTER.

Business will be continued at the same Establishment, by EDWARD SEYMOUR.

Farmington, February 2, 1835. 4w56

Figure 41. Notice of Seymour, Williams & Porter's dissolution appearing in the Connecticut Courant, *February 16, 1835.*

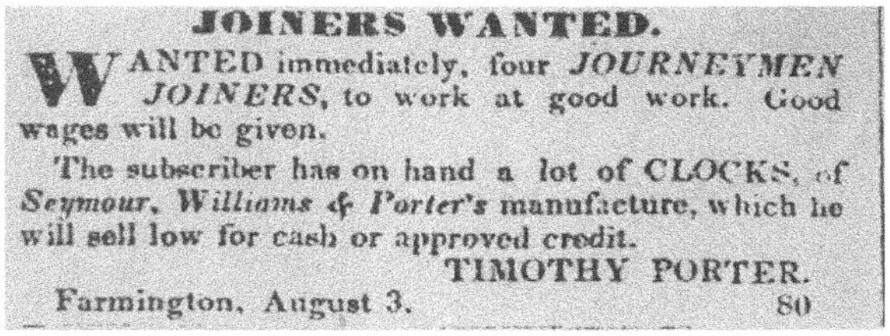

JOINERS WANTED.

WANTED immediately, four *JOURNEYMEN JOINERS*, to work at good work. Good wages will be given.

The subscriber has on hand a lot of CLOCKS, of *Seymour, Williams & Porter's* manufacture, which he will sell low for cash or approved credit.

TIMOTHY PORTER.

Farmington, August 3. 80

Figure 42. Timothy Porter's advertisement appearing in the Connecticut Courant, *August 10, 1835.*

then in Watertown, Connecticut, and then apparently fleeing several creditors and removing to New York City about 1833,[51] where he became the prominent partner in the clock sales firms of Smith & Brothers and Smith's Clock Establishment.[52]

Suppliers to Seymour, Williams & Porter are listed in Table 4, although the accounts provide little detail about them. In a few cases the authors recognized suppliers and inserted the items supplied (in brackets), which are suggested here on the basis of the records of other clock firms or sources. Otherwise, the entries are exactly as they appear in the accounts.

Among persons and firms listed in Table 4, Charles Sigourney, a hardware purveyor of Hartford, was a supplier to other clockmakers. A. & T. Hanks, then of Hartford, were Alpheus and Truman Hanks, brother and son, respectively, of Benjamin Hanks, a clockmaker and bell founder. Alpheus and Truman operated an iron foundry.[53] George Welch & Son were also iron founders, and the son was future clockmaker Elisha N. Welch.[54] Perhaps both the Hankses and George Welch & Son supplied weights.

Table 5 lists individuals and firms found in the Seymour, Williams & Porter accounts who were identified solely through outside sources, such as books and periodicals.

In addition to the above, although the nature of their interactions with Seymour, Williams & Porter cannot be ascertained on the basis of the scant information provided therein, the names of a number of Farmington and/or Unionville residents appear in the pages of the firm's accounts. These include: Ambrose Hart (a builder

Table 4. Suppliers to Seymour, Williams & Porter, with items or services supplied. In the few instances where information about the items supplied was gleaned from records of other clockmakers, the tabulated information from external sources is enclosed in brackets. References are provided for all sources of information outside the Seymour, Williams & Porter accounts.

Name	Item, profession, or professional service supplied	Reference, if from source other than SW&P accounts
Barber	blanks	
A. Bidwell & Co.	glass	
Stephen Bishop	oil, turpentine	
Brainard	stoves	
A. O. Brodie	glass	
P[hilemon] Canfield	printer	
James Church	rope	
Cooley & Smith	glue	
H. Cowles	tables	
T. Cowles	lead	
Fenn	clock keys @ $15.00 per 500 and hinging	
Folsom & Hurlbut	labels	
[A. & T.] Hanks	iron foundry	
F. Hart	use of pattern	
Alfred Hull	painting	
J[oseph] Hurlbut	labels and advertising	
Lee & Butler	[alcohol, acid, etc.]	*Timepiece Journal* Vol. 1, No. 5, p. 81; and Vol. 2, No. 8, p. 177.
Lewis[,] McKee & Co.	[hardware]	*Timepiece Journal* Vol. 2, No. 8, p. 177.
Ludington	faces	
George F. Merriman	oil	
Packet Co.	freight	
Rodney Payne	floor plank	
Enos F. Richards	wagon	
J. J. Roosevelt & Son	glass	
Sam[ue]l Sanford	sleigh	
C[harles] Sigourney	[hardware]	*Timepiece Journal* Vol. 1, No. 5, p. 81.
Steam Boat Co.	glass freight	
B. N. Strong	weights, goods	
Capt. P[omroy] Strong	taxes	
S. Trowbridge	freight	
S. Warren	lumber	
George Welch & Son	weights	
Ozem Woodruff	brick	
[Unidentified]	cases @ $225.00 per 100	

Table 5. Individuals and firms appearing in Seymour, Williams & Porter accounts, whose occupations were identified by means of outside sources.

Name	Occupation	Reference
Isaac D. Bull	Apothecary of Hartford, CT; sold paints and window glass.	Mary Todd, comp. *Thomas & Susannah Bull, Part 2.* (Lake Forest, IL: Heitman Printers, 1983.)
Burnham, G.	Possibly Gordon Burnham, partner in brass mfg. firm of Benedict & Burnham, incorporated in 1834.	Joseph Anderson, ed. *The town and city of Waterbury, Connecticut.* (New Haven, CT: Price & Lee, 1896): 303–304.
Cowles & Co.	Merchants on Farmington Canal.	*Connecticut Courant,* July 21, 1829.
Eli Gilman	Cabinetmaker of Hartford, CT.	*Conn. Hist. Soc. Bull.,* Vol. 33, No. 1 (Jan. 1968.)
Goodwin & Co.	Printers and stationers of Hartford, CT 1825-1854.	D. R. Slaght. *Printers of Hartford 1825-1860,* (unpublished manuscript), NAWCC Library.
William Griswold	Builder of Farmington.	Mabel Hurlburt. *Farmington town clerks and their times.* (Hartford, CT: Finlay Press, 1943): 338.
Hawley & Pickett	Furniture makers of Ridgefield, CT who exported to the South; The Pickett in this partnership was Rufus Pickett, a carver.	*Conn. Hist. Soc. Bull.,* Vol. 33, No. 1 (Jan. 1968.)
Allyn Marsh & Co.	Forwarding merchants of Hartford, CT.	*Connecticut Courant,* September 11, 1832.
Dr. A[sahel] Thompson	Physician of Farmington.	*Farmington, the village of beautiful homes.* (Farmington, CT: A.L. Brandegee and E.N. Smith, 1906.)
Leonard Winship	Cabinetmaker of Farmington.	Christopher Bickford. *Farmington in Connecticut.* (Cannan, NH: Phoenix Publishing, 1983): 248.
Thomas & Joshua Youngs	Partners in Farmington River Manufacturing Co.; dealers in iron, brass, screws, etc., who sold & leased water power sites in Unionville.	Christopher Bickford. *Farmington in Connecticut.* (Cannan, NH: Phoenix Publishing, 1983): 249–250.
William Nevins	Carriage maker of Farmington.	CT Comptroller's Return Schedules, Town of Farmington, CT, 1839. Old Sturbridge Village, MA, microfilm collection. (CT State Library).

of Unionville), Daniel B. Johnson, Noah Porter Jr., Winthrop M. Wadsworth, Marcus Wadsworth, D.[aniel] B. Tuttle, E. O. Gridley, Russel Richards, B. Terrill, N. L. Phelps (a taverner; also the father-in-law of Timothy Porter), Reuben Hawley, Martin Cowles (merchant), and Samuel Deming (merchant). Dr. [Asahel] Thomson, a local physician, received $30.28 in cash on April 1, 1833, for treating Edward Seymour for an unidentified illness. Interestingly, a number of women also appear in the accounts, including: Clarissa Sweetland, Lucy Benham, Lucy Morse, Harriet Warner, Polly Langdon, Clarissa Wadsworth (wife of Sydney Wadsworth), Harriet Hart, Chloe Porter, Cynthia Goodyear, Mary and Paulina Williams, Sally Booth, Julia Ann Root, Orra and Almira Hunt, and "Mary Robberts, Col'd girl." Eunice Thomson appeared in both the accounts, and as the recipient of note in the amount of $160 given to her by the firm on February 6, 1834, found in the aforementioned papers of Austin F. Williams. According to the information provided with the note, the firm paid Thomson by giving her another note. The accounts contain no indication of the work that these individuals might have performed for Seymour, Williams & Porter.

The accounts contain few mentions of clocks per se. However, on January 6, 1832, an entry stated, "To Cash Recd of T. Cowles for clocks 452.84" and another on August 19, 1833, "To 6 clocks 52.50." Additional entries recorded June 26, 1834, are interesting because they document several different types of clocks, with prices, being produced by Seymour, Williams & Porter:

To 4 8 da[y] clocks	9-	36...
To 4 8 da[y] clocks	7½	30...
To 4 8 da[y] clocks	7--	28..
To 12 30 hr clocks carved	5--	60...
To 12 30 hr clocks alarm	5.	60...
To 30 30 hr clocks plain	4½	135...
To 5 Extra finish faces	9c	.63 [note math error]
To Cartage	1.00	350.63[55]

Finally, on January 22, 1835, "To 1 Clock s[o]ld in Hartford [Ct.] 7.00[+]".[56]

Accounts for a few individuals or pairs of individuals need to be mentioned. The first is "Nov 17 [1834] By Cash pd on Amzi Barns Note 50.."[57] Amzi Barns of Burlington, a major creditor of Edward Seymour's, will be encountered again in the next chapter. The following separate items are also noted:

July 30 [1834] By Cash pd Anson Williams on note 10.00

August 21 [1834] By Cash pd Edward Saxton on note 10..

Jany 16 [1835] By Cash pd F. P. Halls note 200..

Dec. 9 [1836] By Cash pd Saxton & Williams. Note & int 97.30

Jany 13 [1837] By Cash pd Edward Seymour on F. P. Halls Note 200..

Jany 13 [1837] By Cash pd Edward Seymour on F. P. Halls Note 5.00

Mar 28 [1837] By Cash pd E. N. Sexton Work pr Bill 12.50

August 20 [1837] By Cash Sextons Note in full 238.00

Oct 23 [1843] By Cash pd Anson Williams on Note 25...[58]

The first three dates are before the dissolution of the firm of Seymour, Williams & Porter; the other six are not. The individuals F. P. Hall, Anson Williams, and Edward Sexton will appear in the next chapter as junior partners, along with Edward Seymour, in the firm of Seymour, Hall & Co., the successor to Seymour, Williams & Porter. Do the last six entries in the above list show that Edward Seymour's new firm, Seymour, Hall & Co., was charging expenses against the old firm, Seymour, Williams & Porter? Probably not. Rather, the first three entries show that these individuals had some relationship with the old company, perhaps as clock factory workers or peddlers. Nonetheless, the last six notes suggest that despite forming a new business in New York

City, Austin F. Williams (and by extension, Williams's father-in-law, Timothy Cowles), continued to quietly underwrite Seymour, Hall & Co.'s finances. Note that four of the notes were given during 1837, the year in which the entire nation's economy collapsed, triggering a severe, prolonged, economic depression. It is against this disastrous backdrop that the remainder of our story is juxtaposed.

Returning now to Court Cases E and F of Table 2A, we find more obvious evidence of financial distress. Seymour, Williams & Porter had given Timothy Cowles a total of four notes dating to before and after the firm's dissolution on February 2, 1835, but had failed to pay them. One of the notes, given to Cowles on January 31, 1834, was for a very large sum of money: $31,000, which was also recorded in the firm's accounts. Peculiarly, Seymour, Williams & Porter had been dissolved for nearly six months when the firm gave Timothy Cowles two additional notes on July 1, 1836. Similarly, the subject of Court Case G was a note Seymour, Williams & Porter gave to the plaintiff in the case, Samuel Deming, on February 25, 1836; as of that date the firm had been dissolved for over one year. An entry dated December 1, 1841, in Seymour, Williams & Porter's accounts, which reads: "Execution in part in favor S. Deming [$]181.65", is likely related to the outcome of Deming's suit. We will return to Court Cases E and F, and the matter of Seymour, Williams & Porter's large unpaid debts to Timothy Cowles, in the next chapter.

Meanwhile, on January 24, 1835, only five days before Timothy Porter sold out his interest in Seymour, Williams & Porter to Seymour and Williams, an incident occurred which seems to bear on the firm's dissolution. On January 26, seven men: Horace Wilcox, George Barber, Anson Williams, Edward Sexton, Daniel B. Johnson, Asa B. Darrow, and Daniel B. Crumb (sic; the last, probably an error, seems to refer to Frederick W. Crum), were summoned to appear before Henry Lewis, a justice of the peace for Hartford County, at Lewis's office in Farmington, on February 2. The reason for the summonses was a complaint brought by Levi B. Rowe "of said Farmington."[59] Because most or all of them appeared in Seymour, Williams & Porter's accounts,[60] both the plaintiff (Rowe), and all of the defendants, seem to have been employees of the firm. All, including Rowe, were in their twenties.

The incident that took place was described in Levi B. Rowe's declaration (complaint), presented to Lemuel Whitman (another justice of the peace), although the act or acts that precipitated it is not. The relevant portion of this document is transcribed as follows:

> ...at said Farmington on or about the 24th day of January 1835 the Defendants did with force and arms an assault make upon the body of the plaintiff and did with like force viz. their hands, fists, feet and knees violently beat and bruise the plaintiff and did with violence throw him upon the floor of a certain school house in sd town where the plaintiff then was, and did then and there trample upon, kick, and bruise the plaintiff and with like force drag him out of the door of said school house into the public highway by which doings of the Defendants the plaintiff was much bruised and injured in his head, shoulders, breast, sides, ribs and private parts, suffered much pain thereby and was put to great expense of surgical aid, his case with papers was knocked from his hand, his pen-knife and pocket book with money therein was thrown from his pocket and some of said money was lost...[61]

The victim, Rowe, claimed $35 worth of damages. Until 1833, Unionville's old schoolhouse had been located in the vicinity of, but across the road from, both the clock factory and Edward Seymour's home. By 1835, it had been relocated a short distance westward, to the later site of the Episcopal Church. According to extant early records of the Unionville School District in the collection of the Farmington Public Library, Levi B. Rowe was not its schoolmaster at the time of the alleged assault.

A number of witnesses were summoned to appear before Justice of the Peace Lewis on February 2 to testify in the case. Although the nature of their testimony is not known, they included: Edward Seymour, Lynde Preston, George Pierpont, Sherman Pierpont, Oliver Moore, Elijah O. Gridley, George Richards, Russel Richards, Daniel Woodruff, Daniel Bradley, Edwin Carrington, and others. Edward Seymour was, of course, the principal partner in Seymour, Williams & Porter. Sherman Pierpont and Oliver Moore were either current or former operators of the nearby Patent Wood Screw Manufactory. George Richards was a Unionville resident whose properties abutted those of several clock factory workers. Elijah O. Gridley and Daniel Woodruff may have been school board members.[62] Russel Richards appeared in Seymour, Williams & Porter's accounts on several occasions beginning in March 1834. Edwin Carrington was a Farmington physician. Presumably, he had treated Rowe's injuries.

If evidence that the men had assaulted Rowe was found, the case could have been referred to Hartford County Superior Court for criminal trial—a very serious matter. However, when all of the parties and witnesses (no attorneys were mentioned) appeared before the justice of the peace court on February 2, Levi B. Rowe withdrew his suit, at which point the defendants petitioned to recover their court costs. The court adjourned until the next day, when "judgment rendered for costs against Plaintiff, & Execution granted."[63] Although insufficient information is available to fully explain the circumstances surrounding the incident and its rather curious resolution, the matter seems to have ended there. No entries related to the court case were found in Seymour, Williams & Porter's accounts. Nonetheless, it is unlikely to have been a coincidence that the firm of Seymour, Williams & Porter dissolved on the very same day that the defendants appeared before Farmington's justice of the peace.

It took a long time to unwind the firm of Seymour, Williams & Porter and its financial affairs. In fact, the latest date encountered in the first part of the firm's accounts was November 1845, and in the second part, May 21, 1850—both far beyond the date of the firm's dissolution on February 2, 1835. We will return to the evidence provided in the firm's accounts later in this narrative, as subsequent developments unfold. Meanwhile, in the next chapter we will return once more to the year 1835, to examine the chaotic events that occurred next, as Edward Seymour, his behind-the-scenes financiers, and others, attempted to fill the void in clock making in Farmington and Unionville after the dissolution of Seymour, Williams & Porter.

Notes and References

1. Kenneth D. Roberts and Snowden Taylor. *Eli Terry and the Connecticut shelf clock.* 2nd ed. Fitzwilliam, NH: Ken Roberts Publishing Co., 1994.

2. Mabel Hurlburt. *Farmington town clerks and their times.* Hartford, CT: Finlay Press, 1943.

3. Christopher Bickford. *Farmington in Connecticut.* Canaan, NH: Phoenix Publishing, 1983.

4. George Dudley Seymour (compiler). *A history of the Seymour family.* New Haven, CT: Tuttle, Morehouse & Taylor Co., 1939.

5. Mabel Hurlburt. *Farmington town clerks,* 1943.

6. Kenneth D. Roberts and Snowden Taylor, *Eli Terry,* 1994, p. 300; and W. R. Brink & Co., *An illustrated historical atlas map of Randolph County, Illinois.* Edwardsville, IL: W.R. Brink & Co., 1875. The biographical sketch of Edward Seymour appearing in the latter source may have been based on an actual interview with him.

7. Ibid. See also: Mabel Hurlburt, *Farmington town clerks,* 1943.

8. Arthur Hughes and Morse Allen. *Connecticut place names.* Hartford, CT: Connecticut Historical Society, 1976.

9. Raymond K. Brooks. "Early manufacturing in Unionville." Unpublished manuscript, October 1976. Farmington Public Library, Local History Center.

10. George Dudley Seymour, *Seymour family,* 1939.

11. Farmington Land Records (FLR), Volume (V.) 44, pp. 228–229, December 30, 1829.

12. FLR V. 44, p. 228, December 30, 1829.

13. FLR V. 44, p. 229, December 30, 1829.

14. FLR V. 32, pp. 529–530, February 18, 1800.

15. FLR V. 44, pp. 324–325, August 14, 1830.

16. FLR V. 44, pp. 515–516, November 27, 1830.

17. "Seymour, Williams & Porter Account Book 1831–1850, Unionville Clocks" [modern designation]. Vols. 1–2. MS91865, Connecticut Historical Society.

18. Latter Day Saints, Family Search website (online source).

19. Calvin Cowles (compiler). *The Cowles families in America.* New Haven, CT: Tuttle, Morehouse & Taylor, 1929. Some additional information about Timothy Cowles was provided in Chapter 1.

20. Connecticut Death and Cemetery Record Index, Connecticut State Library.

21. Oliver S. Phelps (compiler), and Andrew Servin. *The Phelps family of America.* Pittsfield, MA: Eagle Publishing, 1899.

22. Chauncey Jerome. *History of the American clock business for the past sixty years and life of Chauncey Jerome, written by himself.* New Haven, CT: F. C. Dayton Jr., 1860, reprinted by the American Clock & Watch Museum, 2001: 29. See also: Arthur L. Brandegee and Eddy N. Smith. *Farmington, the village of beautiful homes.* Hartford, CT: A.L. Brandegee and E.N. Smith, 1906.

23. Connecticut State Library, Vital Records Index.

24. Oliver S. Phelps and Andrew Servin, *Phelps family,* 1899.

25. Hartford County County Court Files, Box 416, RG3, Connecticut State Library.

26. Ibid.

27. FLR V. 44, p. 360, March 26, 1831. See also: FLR V. 46, p. 1, May 30, 1831.

28. FLR V.44, p. 358, March 26, 1831.

29. FLR V. 44, pp. 546–548, June 11, 1832.

30. FLR V. 45, p. 81, November 9, 1832.

31. "Sanford & Hawley building/Seymour, Williams & Porter clock factory." Historic Resources Inventory, State of Connecticut, Historical Resource Commission (August 1985). This building was still standing in 2014.

32. A few pillar and scroll-style clocks, all in poor condition, bearing Seymour, Williams & Porter labels, are known. The authors are unaware of an intact example.

33. See: Snowden Taylor. "Update on 30-hr. wood Terry-type movements." *NAWCC Bulletin*, No. 237 (August 1985): 473–477; and Snowden Taylor, unpublished update. Seymour, Williams & Porter was not identified as a movement-making firm before the publication of the August 1985 article.

34. Ibid.

35. G. Russell Oechsle and Helen Boyce. *An empire in time: clocks & clockmakers of upstate New York .* Columbia, PA: National Association of Watch & Clock Collectors, 2003.

36. Ibid.

37. Christopher Bickford, *Farmington in Connecticut* (1983).

38. *Connecticut Courant,* July 8, 1833.

39. Ibid. This portion of the toast was also quoted in Christopher Bickford, *Farmington in Connecticut.*

40. FLR V. 46, p. 159, September 25, 1833.

41. FLR V. 45, pp. 138–139, May 2, 1834.

42. Henry A. Castle. *History of Plainville, Connecticut 1640–1918.* New Britain, CT: Hitchcock Printing, 1996.

43. See: New Britain Cemetery inscriptions, Cemetery 1, p. 67; Plymouth Vital Records, V. 1, p. 21; and Bristol First Church Records, V. 2, p. 219; all in the Connecticut State Library.

44. Hartford County County Court, Executions, Box 530, RG3, Connecticut State Library.

45. FLR V. 45, p. 164, January 29, 1835.

46. Arthur L. Brandegee and Eddy N. Smith, *Farmington, the village of beautiful homes,* 1906.

47. Letter dated January 10, 1837, Timothy Porter to Horace Cowles, John Treadwell Papers, RG62:25, Connecticut State Library.

48. *Cog Counter's Journal,* No. 27, p. 22; and Kenneth D. Roberts and Snowden Taylor, 1994, p. 351.

49. "Research Activities & News." *NAWCC Bulletin,* No. 276 (February 1992): 49.

50. Snowden Taylor. "H. Welton & Co., 1839–1846." *Timepiece Journal of the American Clock & Watch Museum,* Vol. 2, No. 8 (Winter 1983): 171–192.

51. Longworth's *New York City Directory 1833–1834* and *1835*, listings for "Smith, R. & H.W."

52. Doggett's *New York City Directory 1841–1842.*

53. W. David Todd and Richard Perlman. "An early factory clock by Benjamin & Truman Hanks." *NAWCC Bulletin* No. 306 (February 1997): 22.

54. Kenneth D. Roberts. *The contributions of Joseph Ives to Connecticut clock technology 1810-1862,* 2nd ed. Fitzwilliam, NH: Ken Roberts Publishing Co., 1988.

55. "Seymour, Williams & Porter Account Book 1831–1850, Unionville Clocks" [modern designation]. Vols. 1–2, MS91865, Connecticut Historical Society.

56. Ibid.

57. Ibid.

58. Ibid.

59. *Rowe vs. Wilcox, Barber, et al.* Town of Farmington Justice of the Peace Records, Box 39c, RG62:52, Connecticut State Library. Note that a riot precipitated by an antislavery lecture held at Farmington on December 15, 1835, appears unrelated to the alleged assault described here, which took place nearly one year previously, in January 1835.

60. It is probable that Horace Wilcox is the "H. Wilcox" who appeared frequently in Seymour, Williams & Porter's accounts.

61. *Rowe vs. Wilcox, Barber, et al.* Town of Farmington Justice of the Peace Records, Box 39c, RG62:52, Connecticut State Library.

62. Mabel Hurlburt, *Farmington town clerks*, 1943.

63. *Rowe vs. Wilcox, Barber, et al.*

CHAPTER 3

Unionville Firms: Edward Seymour and Austin F. Williams; Seymour, Hall & Co.

Following the dissolution of the clock-making firm of Seymour, Williams & Porter on February 2, 1835, no deeds marked the transfer of the firm's land and buildings to a successor firm.[1] However, on August 7, 1835, Austin F. Williams and Edward Seymour, two of the three former partners in Seymour, Williams & Porter, executed the following agreement:

This indenture made the seventh day of August AD 1835, between Austin F. Williams of Farmington, in the County of Hartford, of the one part, and Edward Seymour of said Farmington, of the other part, witnesseth, that the said Williams, for, and in consideration of, the yearly rents and covenants herein after reserved and contained, on the part of the said Seymour...to be paid, observed, and performed, hath demised, granted, and to farm letten, and by these presents, doth demise, grant, and to farm let unto the said Seymour one equal, undivided half of all and singular the lands, with the buildings thereon situated in said Farmington, at a place called Unionville, and such of the tools, implements, apparatus, and machinery belonging thereto, as conveyed to said Williams, by virtue of four deeds, herein after referred to viz.: One deed executed by said Seymour, and recorded in Farmington records Vol. 44, page 547; one executed by Solomon Langdon, recorded Vol. 45, page 81; one executed by said Seymour, recorded Vol. 45, page 139; and one other, executed by Timothy Porter, and recorded Vol. 45, page 164: To have and to hold the premises above described with the appurtenances thereof, unto the said Seymour...from the first day of July AD 1835, for, and during the term of three years thence next ensuing,

and fully to be complete and ended: yielding and paying therefor [sic] yearly, during the said term, unto the said Williams...the yearly rent of Five Hundred dollars, in and upon the first day of July in each year during said term: and if it shall happen that the said yearly rent above reserved, or any part thereof, shall be behind and unpaid by the space of thirty days next after...then, and from thenceforth it shall and may be lawful to and for this said Williams... into the said premises to re-enter and the same to have again, repossess, and enjoy, as in his or their former or first estate right and title... And the said Seymour...doth covenant and grant to and with the said Williams...that he the said Seymour...shall and will well and truly pay, or cause to be paid, unto the said Williams...the said yearly rent above reserved, at the days and times and in manner and form above expressed, and the premises, at the end of said term, unto the said Williams...shall and will peaceably and quietly, leave, and yield up. And the said Williams...doth covenant and grant, to and with the said Seymour...that he the said Seymour... shall and may...under the yearly rent and covenants, herein...peaceably and quietly, have, hold, occupy, possess, and enjoy all and singular [the] tenements and premises above mentioned, with their appurtenances...during the said term hereby granted, without the let [sic], trouble, hindrance, molestation, interruption, and denial of him the said Williams [or] of any other person or persons claiming, or to claim, by, from, through, [or] under him. In witness whereof the persons first above named, have hereunto set their hands and seals, the day and year first above written.

Signed, sealed, and delivered,
in presence of
A.F. Williams   {L.S.}
Edwin Woodruff
Edward Seymour   {L.S.}
Horace Cowles.

Hartford County ss: Farmington August
AD 1835.

Personally appeared Austin F. Williams and
Edward Seymour, signers and sealers of the
foregoing instrument, and acknowledged the
same to be their free act and deed, before me.

Horace Cowles, Justice of Peace.[2]

According to the lease, if the rent was paid on time,
Edward Seymour would be the sole proprietor, half
by ownership, and half by rental, from July 1, 1835,
until the lease expired on July 1, 1838. However, if the
annual rent went unpaid for 30 days or more, Austin
F. Williams was free to "re-enter" or "repossess" the
properties, presumably meaning he could rent them to
someone else.

On September 5, 1835, Seymour and Williams
mortgaged their entire industrial complex to Timothy
Cowles for $9,000.00, to cover a single note in the
same amount dated July 1.[3] In view of Seymour and
Williams's already sizeable debts to Cowles, this was an
astounding sum, and undoubtedly meant that Timothy
Cowles, an influential local merchant and investor then
serving in the state senate,[4] had been financing the firm
of Seymour, Williams & Porter (now dissolved, but with
Seymour and Williams jointly owning the land and facilities), from the beginning. With the breakup of Seymour,
Williams & Porter, Cowles probably wanted the security
of a mortgage. Seymour and Williams, like many other
Americans during the mid-1830s who were embracing
easy credit and a booming national economy (albeit one
based largely on speculation) to invest in manufacturing,
were deeply in debt.

The borrowing continued. On May 10, 1835,
Solomon Langdon passed away.[5] As the executor of his
estate, Farmington attorney Timothy Pitkin had been
directed to sell Langdon's properties at public or private
sales. Like a progressive modern real estate man, Pitkin
divided Langdon's land into lots. During the summer of
1835, he commissioned Daniel Woodruff to produce a

survey map of Solomon Langdon's farm, as subdivided.[6]
A portion of Woodruff's subdivision map is shown in
Figure 43. As depicted, the map points westward toward
the top of the page. This key piece of documentation,
showing the numbered lots into which Langdon's real
estate was divided, allows virtually every lot purchased
by Seymour and Williams, and by Edward Seymour
alone, to be identified.

On September 8, 1835, following an auction of the
subdivided parcels of Langdon's farm remaining unsold
as of September 1,[7] Timothy Pitkin deeded Lots No. 2,
4, 5, 6, and 18 to Seymour and Williams,[8] who were not
partners, but co-owners of the former partnership lands
and facilities. The parcels were generally located in the
midst of the other industrial lands of the former firm of
Seymour, Williams & Porter, and all were described as
being located "at a place called Unionville." Seymour
and Williams paid a total of $1,486.20 for the five lots.
Immediately thereafter, they mortgaged the properties to the Ecclesiastical Society in Farmington for the
same amount.[9] On September 10, 1836, Seymour and
Williams transferred "Lot No. 5 on the Village Square"
to George Barber,[10] a clockmaker or clock factory
worker who appeared often in Seymour, Williams &
Porter's accounts, for only $1.00. Two and one-half
years later, on February 15, 1839, Barber sold this lot to
Edward Seymour's brother-in-law, William Bradley, for
$1,200.[11]

It is hard to see why Williams would want to be
involved jointly with Seymour in additional land purchases after freeing himself from the clock business
by his agreement, above, of August 7, 1835. It would
appear that Edward Seymour, Austin F. Williams, and
Williams's father-in-law, Timothy Cowles, held in common an optimistic vision of the long-term prospects for
both clock making and industrial land in Unionville.

At this point there was no real clock firm or partnership consisting solely of Seymour and Williams. No such
clock labels have been reported. However, author MJD
located two versions of the same hand-drawn period
map that both show the clock factories in Unionville
under discussion. Both are undated, and previously
thought to date to circa the late 1820s, but now obviously 1835 or a little later.[12] The printed version, more
legible, is presented here, as Figure 44. The top of the
map points northward. Note the two buildings each
marked "Clock Factory", and the waterway marked
"Seymour & Williams Canal", running roughly southward, parallel to Roaring Brook (not marked), from

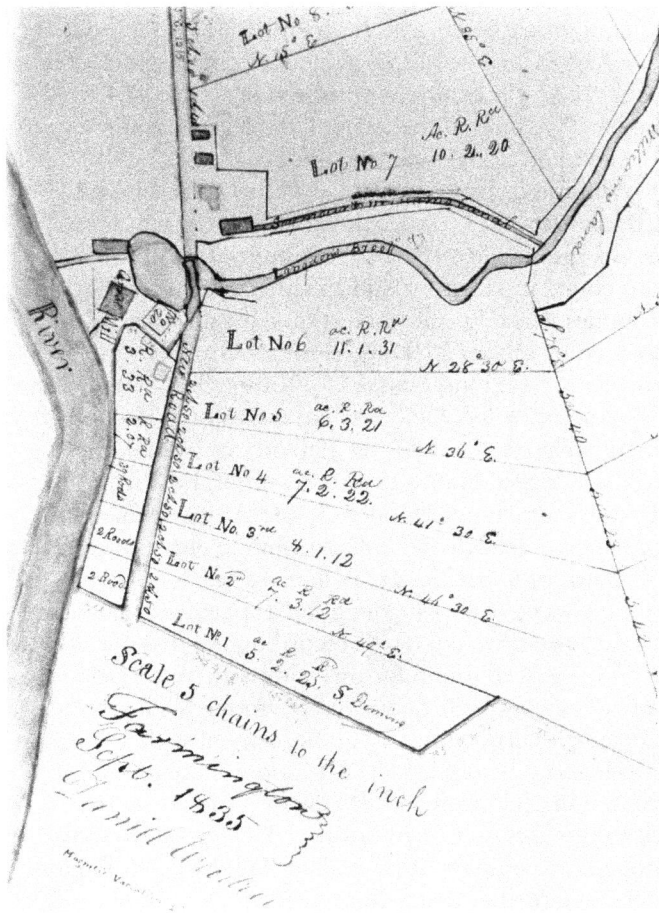

Figure 43. Subdivision map of Solomon Langdon's farm, by Daniel Woodruff, September 1835, from "Maps of Old Farmington," compiled by Julius Gay. The top of the map points westward. (Courtesy of Connecticut State Library.)

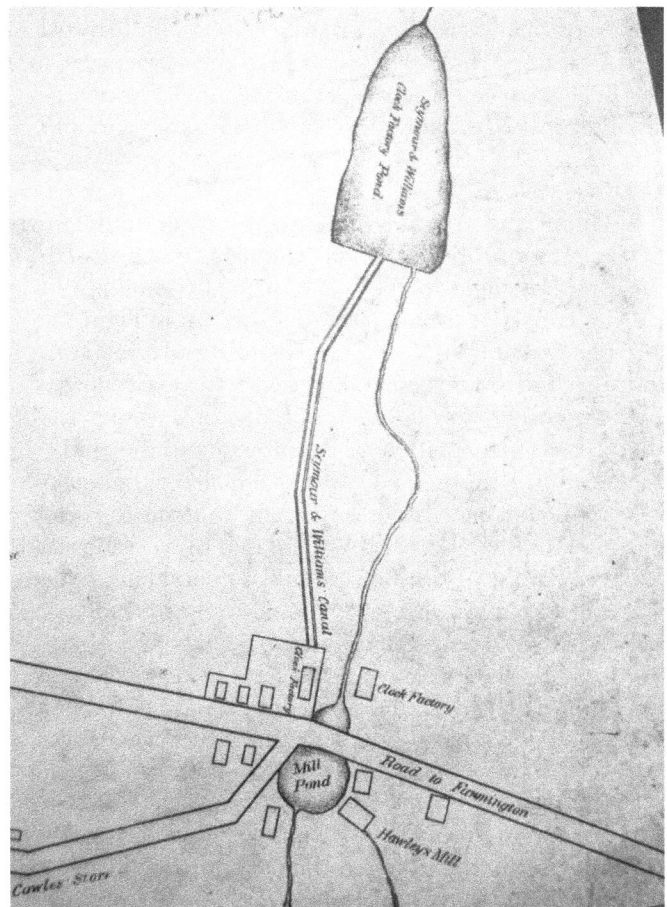

Figure 44. Portion of printed map of Unionville ca. 1835, from "Maps of Old Farmington," compiled by Julius Gay, showing Seymour and Williams's clock factories. (See text.) The top of the map points northward. (Courtesy of Connecticut State Library.)

"Seymour & Williams Clock Factory Pond," down to the western "Clock Factory," to run the wheel, and empty into the "Mill Pond," which in turn empties into the wide, gray region at the bottom of the map, the Farmington River (not named on the map). The factory sawmill was located on the southwest side of the "Mill Pond." Note the "Road to Farmington" that passes over the "Mill Pond" on a bridge, not shown as such. Toward the lower left, partly visible, is "J. & A. Cowles Store." The store sits at the east end of a less wide area (not shown in Figure 44), depicted with black borders above the Farmington River. This was part of the Farmington Canal system, but not clearly defined on the map. This is quite likely the site where "...the feeder dam raised

the water in the river so that canal boats came up the river to the old warehouse built by James and Augustus Cowles,"[13] a convenient arrangement for loading heavy boxes of clocks from the factory. The maps show "Hawley's Mill" to the southeast of the clock factory buildings. This gristmill used water power from Roaring Brook or its millpond at Farmington Avenue, discharging it to the Farmington River.[14] The Hawley property abutted some of the lands Edward Seymour and Austin F. Williams acquired on September 8, 1835.[15]

On September 12, 1835, Edward Seymour alone, apparently eager to speculate in even more land, purchased another portion of the late Solomon Langdon's farm, namely, Lot No. 7, "in Unionville," from Timothy

Pitkin for $217.81.[16] On November 30, 1835, Edward Seymour mortgaged this property, at the same price, to the Ecclesiastical Society of Farmington.[17] The acquisition of Lot No. 7 added about 10 acres to Seymour's already sizeable land holdings near the center of Unionville.

Seymour and Williams appeared together on the town of Farmington's 1835 tax list.[18] The pair was assessed for two manufactories (i.e., each of the two buildings called "Clock Factory" on the map shown in Figure 44), plus 55 acres of land, one sawmill, and a total of four dwelling houses, including two "old yellow houses" built for employees of Solomon Langdon's woolen mill, the "House Occupied by E. Seymour," and the "old Langton [sic]" house. This was essentially all the property of the former firm of Seymour, Williams & Porter, plus the additional property purchased by Seymour and Williams, and by Seymour himself. The total assessment was for $223.50. This may be compared with individual 1837 assessments for Edward Seymour and A. F. Williams, totaling $28.04 and $37.55, respectively, and for Timothy Cowles, whose individual 1837 assessment totaled the whopping sum of $4,784.64.[19]

Table 2B, a continuation of the list of the court cases related to Edward Seymour and his clock-making firms begun in Table 2A in Chapter 2, lists the cases chronologically according to the date on which each was initiated. There is only one court case under Group II, "Edward Seymour and Austin F. Williams." It has already been mentioned, as part of Case F in Chapter 2, with William Bradley as Edward Seymour's "agent." The rather complex circumstances under which Bradley became Seymour's agent, and the court cases, will be discussed in the next section.

Seymour, Hall & Co.

Except for the purchase of the additional industrial property with Williams, the events that took place in 1835 are all consistent with the view that Edward Seymour, whether by choice or not, was establishing an independent clock enterprise using the former Seymour, Williams & Porter property, controlled, but not fully owned by him. Hence it is not entirely surprising that a new clock-making firm, namely, Seymour, Hall & Co., should appear at some point after the dissolution of Seymour, Williams & Porter.

Closely corresponding with Seymour's August 1835 lease of the clock factory properties and equipment, the firm of Seymour, Hall & Co. appeared on the town of

Farmington's tax lists of 1836 and 1837, when it was assessed $150 and $175, respectively, on property that, as of 1837, for example, consisted of a total of four houses.[20] Although the term of the lease was three years, the firm did not appear on later tax lists.

With the commencement of Court Case I in Table 2B, on October 3, 1837, the firm of Seymour, Hall & Co. also began appearing in the records of various courts in Hartford and Litchfield [Connecticut] counties. Case I is the earliest of several court cases located by author MJD listed under Group III, all involving Seymour, Hall & Co. Although the firm name "Seymour, Hall & Co." was not used in the documents related to Case I, the defendants were named as: Frederick P. Hall, Edward Seymour, Edward [N.] Sexton [sometimes "Saxton"], and Anson Williams (no known genealogical relationship to either Austin F. Williams or Cornelius R. Williams). In later court cases these four were clearly identified as partners or members of the firm. No deeds recording the transfer of land have been found for this company. Why? Seymour already owned half of the old Seymour, Williams & Porter properties, and, by his agreement with Williams, was leasing the other half. The additional lands purchased in 1835 were either fully or jointly owned by Seymour. Thus, if Seymour had decided not to make the other members of Seymour, Hall & Co. "full partners," no further deeds were needed.

Court Case I was initiated by Joseph R. Gillett of New Hartford, against the four partners in Seymour, Hall & Co., all of whom resided in Farmington. Consisting of some 40 pages of documents, the case records contribute a wealth of new information to the history of Seymour, Hall & Co.[21] The case began with Gillett's petition "To the Honorable Superior Court for the State of Connecticut to be holden at Litchfield within and for the County of Litchfield on the Third Tuesday of February AD 1838, "Sitting as a Court of Chancery." In his petition Gillett explained:

That on or about the 24th day of July A.D. 1835 your petitioner and Frederick P. Hall, Edward Seymour, Edward Sexton & Anson Williams of Farmington, County of Hartford entered into a copartnership by which it was among other things contracted and agreed between the parties by your petitioner on the one part and the said Hall, Seymour, Sexton & Williams on the other part that the said

Table 2B. Court cases involving Edward Seymour, his partners, and his firms subsequent to the dissolution of Seymour, Williams & Porter.

The date of the writ and summons initiating the court case(s) is provided under "Date of Writ(s)." If the dispute was over non-payment or partial non-payment of a note, the amount of the note is given under "Amt. of Note"; the date of the note is listed in parentheses below. "Outcome / Additional Info" is the outcome of the case as far as known; unless otherwise noted, the outcome is in favor of the plaintiff(s). Additional notes on the case may be found following the outcome in parentheses. The full reference for each court case is provided under "Reference." "Exn." = execution; "plff" = plaintiff; "deft" = defendant; "SH&Co." = Seymour, Hall & Co.; "Default" means the defendants failed to appear, in which case the court rendered judgment in favor of the plffs; "X3" = three lawsuits; "Htfd. Co." = Hartford County; "Ltfd. Co." = Litchfield County"; "Sup." = Superior; "CSL" = Connecticut State Library; "CHS" = Connecticut Historical Society. All amounts are in dollars.

Case Date of Writ(s)	Title of Case	Amt. of Note / (Date of Note)	Outcome / (Additional Info)	Reference
Group II. Edward Seymour and Austin F. Williams				
H 7/29/1839	*Timothy Cowles v. E. Seymour & Austin F. Williams* (See also Case F, Table 2A, Part II.)	550.00 (7/1/1836)	Exn. granted 12/6/1839; defts out of state; may request new trial. (Wm. Bradley, agent.)	Htfd. Co. Co. Court Files, Box 439. RG3, CSL.
Group III. Seymour, Hall & Co.				
I 10/3/1837	*Joseph R. Gillett v. SH & Co.*	(See text.)	See text. Plff awarded 4285.51 damages + costs.	Ltfd. Co. Sup. Court Files, Box 219, RG3, CSL.
J 10/7/1837	*C. D. Cowles v. SH & Co. (X3)*	566.00 (8/21/1837)	Withdrawn 11/1838. (X3)	Htfd. Co. Co. Court Files, Box 436, RG3, CSL.
10/7/1837		1876.46 (3/1/1837)		
10/9/1837		3000.00 (4/1/1837)		
K 9/14/1838 & 7/25/1839	*Asa B. Darrow v. SH & Co.*	200.00 (5/2/1836)	Default; Exn. 11/14/1839. (Marcus Wadsworth, Daniel B. Johnson, dba Wadsworth & Johnson, agents. See text.)	Htfd. Co. Co. Court Files, Boxes 437 & 439, RG3, CSL.
L 3/12/1839	*Williams, Orton, Prestons & Co. assigned to Wm. Beach v. SH & Co.*	100.00 (2/1/1838)	Default; Exn. 4/5/1839. (Note assigned to Beach 8/7/1838.)	Htfd. Co. Co. Court Files, Box 437, RG3, CSL.
M 10/15/1839 (X2)	*Charles I. Langdon, assigned to Wm. Beach v. SH & Co.*	189.24 (2/24/1837) 71.54 (10/31/1837)	Default; Exn. 4/9/1840 (X2); (Notes assigned to Beach; Daniel Bates, agent; defts out of state, may request new trial. See text.)	Htfd. Co. Co. Court Files, Boxes 437, 439, and 440 RG3, CSL.
N 10/15/1839	*Joshua B. Brewer, assigned to Wm. Beach vs. SH & Co.*	398.70 (5/15/1838)	Default; Exn. 4//9/1840. (Defts out of state, may request new trial; Daniel Bates, agent.)	Htfd. Co. Co. Court Files, Box 439, RG3, CSL.

Frederick P. Hall & your petitioner should go on to the State of Alabama & there to receive of the Said Seymour Hall Sexton and Williams Brass & Wood Clocks there to be Sold by the Said Hall and the petitioner and also such other property as said Seymour & others should forward or Ship to the said Hall & the petitioner for that said copartnership should continue for the period of two seasons or Twenty two months...and that said business should be carried on under the name & firm of Gillitt & Hall [sic]...[22]

Under the partners' agreement, a written copy of which was unavailable to be presented to the Court as evidence, Gillett claimed that he was entitled to half the profits and would sustain half the losses, and that Hall, Seymour, Sexton, and Williams were entitled to the other half.

Gillett related that he and Hall proceeded to Alabama, "...took on with them a number of hands employed to sell said Clocks & other property, and continued then in said States of Georgia and Alabama selling said Clocks and other property..."[23] Gillett also stated that he transmitted large sums of money to Seymour, "to wit Eight Thousand dollars" and that Hall "received for clocks and other property so[ld] by him and your petitioner...large sums of money to wit five Thousand dollars..." and that the firm received notes totaling $17,000. Gillett alleged that, upon the expiration of the copartnership, he returned to New Hartford to settle with Hall, Seymour, Sexton, and Williams, bringing with him notes for about $7,000, but "Hall and Seymour deceitfully, fraudulently & forcibly seized upon all the Notes and all other papers..." and "that no settlement touching the accounts of said copartnership has ever been had..." Furthermore, Gillett stated that a just settlement to him would exceed $12,000. Gillett signed his petition at Winchester on October 3, 1837.[24]

Following its hearing on Gillett's petition, the court commanded the sheriff of Hartford County to attach the goods or estate of partners Seymour, Hall, Sexton, and Williams, to the amount of $30,000. Shortly thereafter, on October 3, the sheriff went out and attached property, much of it recognizable as the properties purchased by Edward Seymour in 1835. It is interesting that the same properties were again placed under attachment on October 7, 1837, in Timothy Cowles's suit against Seymour, Williams & Porter for

non-payment of the $30,000 note the firm had given him on January 31, 1834 (Court Case E, Table 2A, Chapter 2). Of course, Timothy Cowles also held a mortgage on the properties that secured his interest in them.

All of the above material was presented to the Court in February 1838. While under the Court's consideration, Gillett added to his petition, as follows:

...your petitioner says that the said Edward Seymour & the said Frederick P Hall have each of them a large amount of Notes, monies & property in their hands to wit fifteen Thousand dollars, being the effects of said Company that the said Seymour, Hall, Sexton and Williams wholly refuse to settle and adjust the concerns of said copartnership and are appropriating the monies & effects of said company to their own use and benefit, and your petitioner further says that the said Seymour & Hall are wholly insolvent and unable to pay their debts and your petitioner verily believes & has good reason to believe that they will dispose of all the effects of said company to their own use - And the petitioner further says that said Seymour & Hall are fraudulently combining & confederating to cheat deceive and defraud your petitioner, and to deprive him of all his right & interest to and in the effects of said copartnership...[25]

Gillett continued, saying he was without adequate remedy at law, and therefore prayed the Court to take his case because he was without common law proof. He prayed the Court to summon Seymour and Hall to appear, asking that they be required to reveal under oath their knowledge of the case, and bring into Court all the papers concerning it. Finally, Gillett asked the Court to enjoin Seymour and Hall not to dispose of or convert any papers or collect on any notes.

Subsequently, the Court commanded the sheriff to summon Edward Seymour, Frederick P. Hall, Edward Sexton, and Anson Williams to appear before the Court on the third Tuesday of August 1838. The Court also issued the injunction requested by Gillett, dated March 9, 1838, which was delivered to the respondents.

There is a gap in the court records, but it is clear that the Superior Court (acting as a Court of Chancery), proceeding in the case of *Gillett vs. Seymour, Hall, and others*, appointed a Committee consisting of Morris

Woodruff and Seth Terry. On July 8 and July 27, 1839, the Committee contacted the plaintiff and the four respondents to set up a meeting. The Committee prepared a report dated October 4, 1839, that indicated the case was "continued from August Term AD 1839 to February Term AD 1840." In its report the Committee concluded that the respondents owed the petitioner $6,126.39 and noted the respondents did not attend the meeting.[26] Certain preliminary documents indicated that the Committee would meet on July 29, 1839; moved to August 5; moved to August 12; but indicated on the Committee report itself as October 3, 1839. Based on information provided in the records of several of the court cases listed in Table 2B (Case H, for example), as of July 29, 1839, Edward Seymour was absent from the State of Connecticut. However, when the February 1840 term of the Court convened, the respondents appeared, and voiced their objections to the Committee's report. They also claimed that the Court, acting as a Court of Chancery, had no jurisdiction in the matter.

The next documents in the files of Court Case I pertained to the Court's February term AD 1842. The gap was partly explained in the Court's summary of events, as follows: "...said cause was duly continued, and reserved for the advice of the Supreme court of Errors & by continuances came to the present term..."[27] In addition, the document stated that Gillett, the petitioner, "comes into court & remits all claims growing out of the following item[s] in the account...by the Committee (to wit) One slave, Horses, Waggons, clocks, harness & divers other articles of visible personal property in Alabama - $4600 -."[28] The Court introduced the change into the amount recommended by the Committee, awarding Gillett $4,285.51 plus court costs of $95.53. Indeed, Gillett's remittance of his claim to the copartners' "visible personal property" in Alabama followed a determination by the State of Connecticut's Supreme Court of Errors that that portion of the copartnership property had never been in the defendants' possession or control, so the Committee had erred in allowing the $4,600 charge. Consequently, the high court ordered the charge "expunged" from the Committee's account.[29]

There is another curious item in the case file. During the 1842 Court term, O. S. Seymour was both Clerk of the Court and Joseph R. Gillett's attorney. A mysterious letter addressed to Mr. Origin Seymour, Litchfield, from Acantha Gillett (relationship, if any, to Joseph R. Gillett unknown at the present time), urged that the documents

containing the claims be preserved in case they were needed during an unidentified bankruptcy proceeding. In fact, Edward Seymour's bankruptcy was already in the works. Origin Storrs Seymour, a prominent local attorney, was later appointed a judge in Litchfield County in 1855. Still later, in 1873, he would become Chief Justice of the State of Connecticut Supreme Court.[30]

One of the documents in the court case of *Joseph R. Gillett vs. Edward Seymour et al.*, just considered, was signed "Harlow Gillett/ an indifferent person." There was an entirely separate court case (not part of Table 2B), *Harlow Gillet of New Hartford vs. Joseph R. Gillett and Frederick P. Hall*, "...both late dealers in Company & Selling Clocks & Boots under the name & firm of Gillet & Hall...", commencing with the plaintiff's complaint and subsequent issuance of a writ of attachment on October 5, 1837.[31] The sheriff went out on October 7 and "attached as the Estate of the Defts...9 pieces of Land and the Buildings thereon standing situated in said town of Farmington..."[32] There follows a description of nine numbered lots, recognizable as being located in and around the heart of Seymour and Williams's industrial land centered on Roaring Brook: "For a more particular description of the above described Land reference to be had to a survey of the Langdon Farm, made by Daniel Woodruff, Esq-which survey [was] lodged in the Clerks office of said Farmington."[33]

The Court eventually rendered judgment and execution thereof in favor of Harlow Gillett, against the former partners in Gillett & Hall, in the amount of $756.92 plus court costs of $30.49.[34] However, between October 27 and December 6, 1838, a constable, and then a sheriff:

> ...made search for Goods & Chattels of the debtors [Gillett and Hall] in this Exn on which to levy to satisfy the same and my fees but could not find any within my precincts Neither could I find the Debtors or Either of them within my Precincts to take their bodies...and I continued so to search for Goods & Chattels, and the Bodies of Debtors...but not being able by the most diligent search through my precincts to find...them - I return this Exn wholly unsatisfied[.]"[35]

An additional court case (also not listed in Table 2B), that of *Joseph Gillet* [no middle initial] *vs. Joseph R. Gillet* [note presence of middle initial "R."] *and Frederick P. Hall,* commenced on October 6, 1837. This was only one day after the commencement of the previously described case, in which Harlow Gillet sued the same two defendants, Gillett and Hall. It is interesting that Harlow Gillet also figured in the suit brought by Joseph Gillet, by acting as the plaintiff's agent. Furthermore, the same sheriff went out on the same date, October 7, 1837, and attached the same nine parcels of land as the "Estate of the Defts," describing the land in the same manner.[36] In both cases the descriptions of the parcels subjected to attachment, including their lot numbers, correspond exactly to parcels owned by Seymour and Williams that were formerly part of Solomon Langdon's farm as illustrated in Woodruff's survey. Both cases reached the April 1838 term of Litchfield County County Court, although the case brought by Joseph Gillet was continued through several additional terms, until October 1838, when the court judged in favor of the plaintiff, finding that Gillet should recover of the defendants the amount of $54.54 in damages and $31.67 costs of suit.[37] Whether execution of the court's judgment was met with a similar lack of success as in the case against Joseph R. Gillett and Frederick P. Hall brought by Harlow Gillet, acting on his own behalf, is unknown.

Returning again to Table 2B, there are five additional court cases (or groups of cases) in Group III, involving partners Edward Seymour, Frederick Hall, Anson Williams, and Edward Sexton [sometimes "Saxton"], the latter two constituting the "& Co." in Seymour, Hall & Co. This firm was mentioned in Reference 1 (p. 302), but except for Edward Seymour, the other partners' names were then unknown. Although the additional cases agreed on the firm's name and the names of its partners, descriptions of the firm varied somewhat from case to case; for example, in two cases the partners were referred to as "manufacturers," and in one case they were described as "dealers." Taken together, all this seems to denote a normal clock-making firm of the period.

In Cases H, M, and N of Table 2B, judgments against the defendants (the partners in Seymour, Hall & Co.) dating from between late 1839 and early 1840 were bonded, in the event the defendants requested new trials because they were convicted in absentia, being absent from Connecticut at the time of the trials. Presumably, they were peddling or collecting on notes. Whether the injunction of March 9, 1838, against collecting the notes that had been given to the firm of Gillet & Hall

for clock sales in Alabama had been lifted by that time is unknown.

In Court Case K, Asa B. Darrow of Unionville, likely one of Seymour, Hall & Co.'s employees, brought suit against the partners in the firm, who were described as "now gone to parts unknown out of this state." Sent out on July 25, 1839, to attach $300 worth of the defendants' property, a sheriff's deputy reported:

> Then by Virtue hereof and by the special direct. of the Creditor, I attached all the right title and interest of Edward Seymour in a certain Dwelling House out building, and Land situated on the main Street in sd Farmington, bounded North & East on John Thompson South on W. S Nevins West - Highway. Also one large two Horse Wagon with 4 Seats, 1 do 2 Horse, 1 Running part for one Horse Shay [sic]. 2 Large Mahogany Logs. 500 Clock Colums. One Circular Saw. 500 Frets. 1 Saw & Frame. 120 Hemlock Boards. About 1000 Cherry Tops & Bottoms. 400 Top Blocks. 30 Carved Mahogany Colums. Lot of Cherry Wheel Stuff. Lot of pinion Stuff. about 5 Cords Sawed Slabs. 1 Lot Timber for House frame. 1 Lot Joist & Rafters. 1 Pile pine Siding 1 Lot Mahogany blocks...[38]

The Court awarded the plaintiff (Darrow) $240.50 in damages and $11.33 costs of suit. At first glance, the list of attached items suggests the discovery of a new clock-manufacturing establishment in Farmington village, but the list is obviously incomplete for a clock factory. However, on the basis of information provided in the Town of Farmington Land Records,[39] the "Dwelling House" referred to in the attachment, where the material was stored or perhaps hidden, was that of Edward Seymour's father, Luther Seymour, who had died in December 1815. As late as 1843, the property in question, located in Farmington village about three miles east of the clock factory site in Unionville, still belonged to Luther Seymour's heirs. Asa B. Darrow's appearance in Seymour, Williams & Porter's accounts was described in Chapter 2.

It is interesting that the focus of Court Case L was a note Seymour, Hall & Co. had given to the clock-making firm of Williams, Orton, Prestons & Co. on February 1, 1838, promising to pay the latter firm $100 at the Phoenix Bank in Hartford. Williams,

Orton, Prestons & Co., in turn, ordered the note to be paid to William Beach, a clock dealer, so Seymour, Hall & Co. became liable to pay Beach. There was no national currency until after the Civil War, and such transactions between parties, while commonplace, not infrequently became the subjects of lawsuits. On March 12, 1839, a sheriff's deputy read the writ and summons to Edward Seymour. However, after serving the writ on Seymour, the deputy noted in his return to the Court that no goods or estate belonging to the defendants could be found on which to make an attachment, "the other defendants having no known residue, agents, or attorney in this state," again providing information on the absence of Seymour's partners. Similarly, Cases M and N centered on notes given by Seymour, Hall & Co. that had been assigned to the plaintiff, William Beach. As early as November 24, 1834, Beach's advertisements had begun appearing in the *Connecticut Courant,* in which he offered clocks "...of different materials, of the best workmanship" for sale to "Pedlars, and others..." from his store on Mill Street in Hartford.[40]

By October 15, 1839, according to the writ of attachment in Court Case M, the defendants, Seymour, Hall & Co., were described as "late manufacturers...now absent and out of state." Daniel Bates of Winchester was served with a copy of the writ as the firm's agent. Charles I. Langdon had ordered a note Seymour, Hall & Co. gave him on February 24, 1837, for $189.24, to be paid to Bates, who in turn ordered it to be paid to William Beach, then "of Hartland," only ten days before the date of the writ. Seymour, Hall & Co. gave Langdon a second note for $71.54, on October 31, 1837. Both notes had gone unpaid.

The plaintiff in Court Case N was cabinetmaker Joshua B. Brewer, of Unionville, perhaps a clock case-maker at the time of the suit. By1844 Brewer became Lambert Hitchcock's partner in a furniture-making venture at Unionville.[41] This was the same Lambert Hitchcock who popularized bronze-stenciled chairs at Barkhamsted, CT, beginning in the 1820s.

Because Frederick P. Hall was a partner in Seymour, Hall & Co. throughout the firm's existence, it is interesting that an "F.P. Hall" between the ages of 30 and 40 years was residing in Tuscaloosa County, AL, at the time of the 1840 U.S. Census. He was still living in Tuscaloosa at the time of the 1850 Census, when his occupation was that of farmer. This may be the same F. P. Hall who died in Connecticut on February 1, 1855, at the age of 45.[42]

Edward [N.] Sexton {sometimes "Saxton"], another partner in Seymour, Hall & Co. may have been injured when the steamboat *Moselle,* bound for St. Louis, exploded near Cincinnati, OH, on April 27, 1838.[43] Apparently, Sexton survived the blast, and returned to Connecticut, where he was identified as a Farmington resident in the 1840 U.S. Census. As of 1850, Sexton was living in Bristol, and still making clock cases.[44]

Considered cumulatively, the data show that Seymour, Hall & Co. had been a clock-manufacturing company that produced and sold clocks and movements. It is not surprising that the firm's 30-hr. wooden movements evolved from those of its predecessor, Seymour, Williams & Porter. Figures 45A and 45B show a half-column and splat-style, Seymour, Hall & Co., Unionville, 30-hr. shelf clock. Its label, like the labels of Seymour, Williams & Porter, credits Eli Terry as the movement's inventor. Its Type 1.411 movement closely resembles the "later" Seymour, Williams & Porter & Co. movements with rounded escape wheel bridge tips (but not fully round).

A second example of a clock by Seymour, Hall & Co. is illustrated in Figures 46A–46C. In addition to crediting Eli Terry, its label states that the clock's movement is brass bushed. However, a small brass bushing visible on the strike side winding arbor in the close-up (Figure 46C) of the clock's 30-hr. movement, also Type 1.411,[45] was probably installed later, as a repair. An interesting variant of this 30-hr. movement, Type 1.412, is known, in which the count wheel pinion bridge resembles the escape wheel bridge in its construction and arrangement on the front plate.[45]

Another 30-hr. movement from a Seymour, Hall & Co. shelf clock is shown in Figure 47. Its design has progressed to where the "figure 8" access aperture has a larger center hole than the two end holes, and the movement is designated Type 1.42.[46] Seymour, Hall & Co. also produced an 8-day wooden movement more like the second generation of 8-day wooden movements produced by Seth Thomas than those of Seymour, Williams & Porter.[47] An example is shown in Figures 48A–48C.[48] A 30-hr. Seymour, Hall & Co. clock in an unusual case, with a time, strike, and external alarm movement much like the alarms produced by Seymour, Williams, & Porter (illustrated in Chapter 2), is also known. Its external alarm movement has mahogany plates (see Figures 49A–49C). The brass corners on the case door front and stenciled tablet (Figure 49A) are not original.

Only days after the clock factory lands and buildings were placed under attachment in the lawsuits brought

Figure 45A. Seymour, Hall & Co., Unionville, 30-hr. wood movement shelf clock, exterior view.

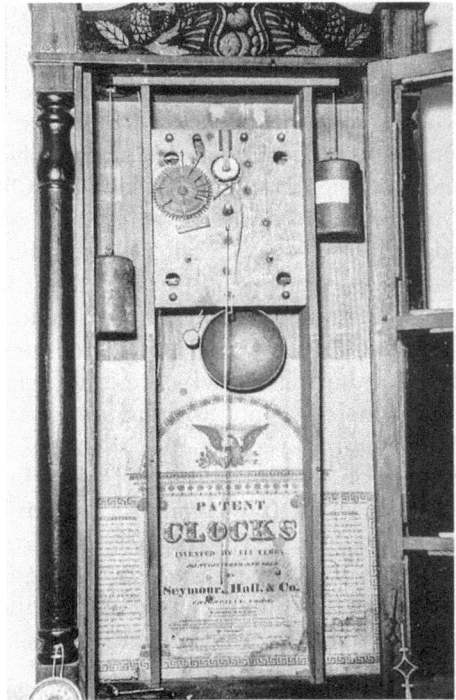

Figure 45B. Clock shown in Figure 45A, door open, showing label and Type 1.411 wood movement by Seymour, Hall & Co. (Figures 45A–45B courtesy of American Clock & Watch Museum.)

Figure 46A. Another example of a Seymour, Hall & Co., Unionville, 30-hr. shelf clock, exterior view.

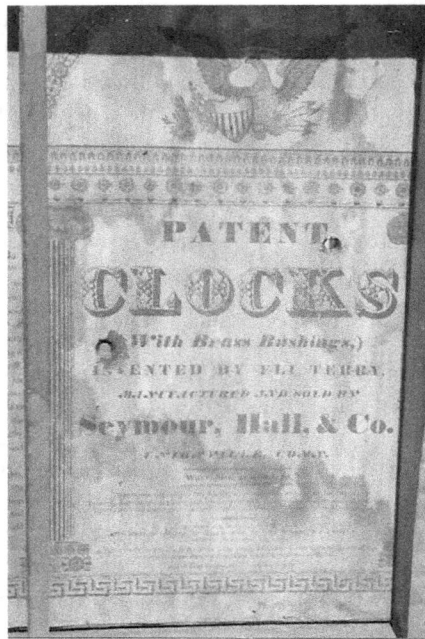

Figure 46B. Label of Seymour, Hall & Co. shelf clock shown in Figure 46A.

Figure 46C. Closeup of Type 1.411 movement of clock shown in Figures 46A and 46B. (Figures 46A–46C courtesy of Craig and Jauna Siebold.)

by Joseph R. Gillett, Harlow Gillet, Joseph Gillet, and Timothy Cowles, by means of a deed dated October 9, 1837, Seymour Hall & Co. subleased (no transfer of land) their manufacturing properties to Chauncey D. Cowles, the 25-year-old son of Timothy Cowles[49]:

> In consideration of one dollar...we do hereby lease unto him, and by this act give unto him possession of all the buildings, viz: the two factories, the saw mill two barns, and kill-dry [sic, kiln-dry], which we occupy, in the prosecution of our manufacturing in Unionville, and which we hold under a lease from Williams and another, for which the said Cowles is to pay us a reasonable rent, and is to have and hold the premises for the space of twelve months from this day.[50]

Lessor Williams was Austin F. Williams; the "another" was, of course, Edward Seymour.

Returning to Table 2B, in Case J, Chauncey D. Cowles brought suit against the firm of Seymour, Hall & Co. for failing to completely pay three notes (dated March 1, April 1, and August 21, 1837, respectively), on October 7, 1837, just two days prior to the date of the abovementioned lease. On October 7, Farmington constable Pomroy Strong reported:

> By virtue of this writ & by direction of the Plaintiff within named I attached as the Property of Seymor[sic], Hall & Co., the following described property to wit, four horses, seven harnesses, three double waggans, One single do, one chaise, four hundred eight day Clock movements, four hundred eight day cases do - 167 thirty hour Clocks - 20 eight day Clocks _ 170 gilt tops - 187 Clock cases - 135 pair eight day weights 2000 Clock faces - 72,000 [12,000?] inch brads - 28 thousand 7/8 inch do - 8 thousand 1 1/8 inch do - 88 alarms - 900 lbs. iron wire - 1 drill lathe & turning wheel - 1 Iron vice - 5 work benches with all fixtures thereto belonging - 1 Time piece in the window - 400 sett wheel stuff - 1800 Clock plates - 6 turning lathes & benches - 1 wheel lathe - 7 drills - Button press - 1 wheel engine - 1 Pinion engine - 7 Circular saws & frames - 1 grind stone & frame - with all the wheels, shafts, pulies [sic] & hangers & all the tools & implements thereto belonging to set the foregoing machinery in

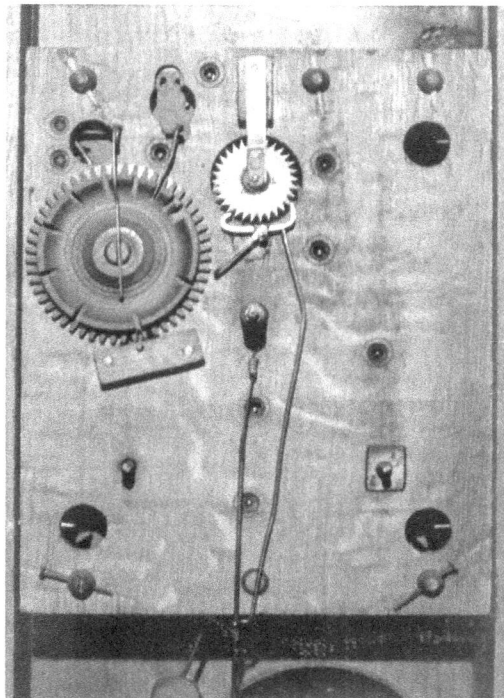

Figure 47. Closeup of "late-style" movement, Type 1.42, made by Seymour, Hall & Co. Exterior image of clock not available. (Ward Francillon photo.)

operation - 100 Tops not covered - 1 screw plate & dies - 3 work benches & screws - 99 Tablets different sizes - 532 Tablets 10 by 19 - 100 looking glass plates - 150 in by eight & seven by nine window glass - 51 lb. No. 1 glue - 259 Pointed glass in frames - 188 picture frames - 8 boxes 10 by 15[?] glass - 7 do - 11 by 11 do - 203 covered Tops - 360 Carved colum[ns] - 836 bronzed Tops - 960 painted Columns not polished - 1090 Columns gilt bronzed & white - 458 setts Column Blocks - 110 pr Top- Bottoms - 379 pair Top Blocks - 617 sash for Clock cases - 575 pair sides for do - 250 feet Mahogany crotcht at the mill - 1177 Tops & bottoms not finished & all parts & parcels of clocks finished & unfinished in East & West factory - 9000 feet Cherry lumber in the saw mill - Lott of Cherry & pine lumber in Langdon Barn - all the mahogany & pine lumber in the saw mill & in the mill yard - & all the water power & all the privileges & appertinances [sic] belonging [thereunto]...[51]

Figure 48A. Seymour, Hall & Co. 8-day wood movement shelf clock, exterior view.

Figure 48B. Seymour, Hall & Co., Unionville, label of clock shown in Figure 48A.

Figure 48C. Seymour, Hall & Co. 8-day wood movement in clock shown in Figures 48A and 48B. (Figures 48A–48C courtesy of Chris Brown; Gene Gissin photos.)

Figure 49A. Seymour Hall & Co. 30-hr. wood movement shelf clock with time, strike, and external alarm movement, exterior view. The corniced case is unusual.

Figure 49B. View of label and exterior alarm in Seymour, Hall & Co. clock shown in Figure 49A.

Figure 49C. Thirty-hr. time, strike, and external alarm movement arrangement in clock shown in Figures 49A and 49B. (Figures 49A–49C courtesy of a private collection.)

Note the presence of both 30-hr. and 8-day clocks and clock cases (undoubtedly with wooden movements) in progress in the "East and West" factories; the pulleys and shafts for water-powered machinery; the wheel and pinion engines; and also the "Time piece in the window." Following the attachment, without access to its tools, stock, and work in progress, it is hard to see how Seymour, Hall & Co. could have continued to produce clocks. Although C. D. Cowles withdrew the suit during the County Court's November 1838 term, the information in the case file sheds no light on how the debt was in fact resolved.

Reference 1 (p. 302) referred to a serious fire occurring in 1837, which was reported in additional (secondary) sources, but all mentions for the fire's date were very vague. Thanks to Clifford Alderman (at the time, president of the Unionville Museum), primary sources documenting the fire's occurrence were found. On April 4, 1838, the *Hartford Daily Courant* reported the following disastrous news (see also Figure 50):

> FIRE AT UNIONVILLE, FARMINGTON.--
> The Saw Factory and Clock Factory owned by Timothy Cowles, and occupied by Seymour Hall & Co. and C.D. Cowles, were destroyed by fire on Monday afternoon last. Nothing of value was saved. Loss estimated from 35 to $40,000--$15,000 insured.

The fire, which had occurred on April 2, 1838, was also reported in H.[orace] Greeley & Co.'s publication *The New Yorker*[52] on April 21. *The New Yorker*'s editors, whose reformist and antislavery views were rapidly gaining influence, could not resist the opportunity to pun, noting: "A wooden clock factory at Unionville, Ct. was destroyed by fire on the 2d inst. Estimated loss $40,000, besides, doubtless, a great loss of *time*."[53]

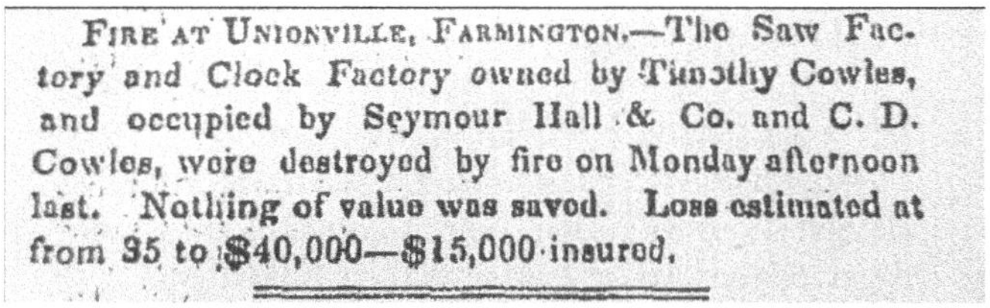
FIRE AT UNIONVILLE, FARMINGTON.—The Saw Factory and Clock Factory owned by Timothy Cowles, and occupied by Seymour Hall & Co. and C. D. Cowles, were destroyed by fire on Monday afternoon last. Nothing of value was saved. Loss estimated at from 35 to $40,000—$15,000 insured.

Figure 50. Report of Seymour, Hall & Co. clock factory fire, Hartford Daily Courant, April 4, 1838.

It seems appropriate to ask the question, "Did the fire on April 2, 1838, bring clock manufacturing to an end in Unionville, within the properties of former or present owners Seymour Williams & Porter; Seymour and Williams; Seymour, Hall & Co.; or Edward Seymour?" Considering the evidence provided in the records of the court case, *Chauncey D. Cowles vs. Seymour Hall & Co.*, in which the firm's clocks in progress, parts, tools, and machinery had all been placed under attachment pending the outcome of the suit (which was not withdrawn until November 1838), and the "deed" of October 9, 1837 (quoted above), in which Seymour, Hall & Co. turned over to C. D. Cowles the entire use of the property for one year, it is likely Seymour, Hall & Co.'s clock making ended at that time. Chauncey D. Cowles is not known to have been a clockmaker, and the "two factories" were the only clock-making sites within the properties of Seymour, Hall & Co.'s predecessor firm, Seymour, Williams & Porter. The latter firm had been dissolved, but the property continued to be owned by Edward Seymour and Austin F. Williams, albeit mortgaged to Timothy Cowles. In the meantime, the Panic of 1837 had brought virtually all forms of business in the young United States to a standstill. So on the face of it, clock making at the Roaring Brook facilities must have ended *before* the fire occurred on April 2, 1838. However, it was not quite the end, as we shall see.

According to the *Courant* of April 4, 1838, the "Saw Factory and Clock Factory [singular]" destroyed in the fire were "owned by Timothy Cowles." The authors initially believed that these facilities were not owned by Cowles until the execution in Cowles's suit against Seymour, Williams & Porter, which execution was variously dated in the court records as April 23, 1838, then "No. 2 June 30, 1838" (Case E, Table 2A, in Chapter 2), and then August 16, 1838. But it may have been well known that Cowles won the judgment against the three partners on March 4, 1838, in the amount of some $16,000. *The New Yorker* also referred to "a clock factory [in the singular]" (see above). The authors present evidence below (and in the next chapter) that suggests that only the western clock factory building shown in Figure 44 (in which the map points westward, toward the top of the page), probably the movement shop, was destroyed in the fire, and that after C. D. Cowles withdrew his suit against the partners in Seymour, Hall & Co. in November 1838, some limited casemaking and/or clock assembly continued within a portion of the eastern factory building until late 1839, albeit under the auspices of others.

On May 18, 1838, six weeks after the fire, Edward Seymour deeded to his brother-in-law and "agent," William Bradley of Farmington (mentioned previously in connection with Court Cases F (Table 2A, Chapter 2), and H (Table 2B, above), one-fourth of an acre of land, for $225.[54] The parcel was "situated in Unionville...bounded southerly and Westerly on land of Timothy Porter, Northerly on Mill Pond, Easterly on highway...with a barn and Cow House standing thereon." Born in Plymouth on May 27, 1806, William Bradley was the son of Chauncey and Betsey (Avery) Bradley;[55] his wife Adaline (Johnson)'s sister Harriet was the wife of Edward Seymour.[56] The barn in question had formerly belonged to Solomon Langdon.

At least one example of a 30-hr. shelf clock labeled: "WILLIAM BRADLEY / (Successor to Seymour, Hall & Co.) / ST. LOUIS, MO." is known (Figures 51A–51C), and it has been suggested that Edward Seymour's firms maintained a depot in St. Louis (Reference 1, page 302). Like the clocks of both Seymour, Williams & Porter and Seymour, Hall & Co., this clock's label credits Eli Terry as its inventor. The maker of the clock's movement (Type 8.2 [unpublished], Figure 51C) is unidentified. Although it is difficult to see in Figure 51C, the slightly rounded tip of the movement's escape wheel bridge is a feature associated with clocks of Edward Seymour's firms. A few other clocks with this movement have been encountered, at least one bearing the label of Charles Mann, Springfield, M[..., Massachusetts?], pasted over Seymour, Hall & Co.'s label, suggesting that perhaps it is a hybrid, made from leftovers plus some parts purchased after the fire. William Bradley, who was also mentioned above in connection with a parcel of land, probably a house lot, deeded him by George Barber on December 10, 1836, leased an additional parcel in the vicinity of the former Seymour, Hall & Co. clock factory, to Timothy Cowles on December 28, 1839.[57] These findings suggest that William Bradley quietly managed the firm's unfinished clock business, perhaps beginning in 1838, until as late as the fall or winter of 1839.

The Aftermath of Edward Seymour's Clock Enterprises

The case of *Timothy Cowles vs. Seymour, Williams & Porter* (Case E, Table 2A, Chapter 2), with its related documentation, contributed some of the most important information toward unraveling the history of the Unionville clock firms. At issue was the note the firm

had given Timothy Cowles on January 31, 1834, promising to pay him $31,000 "On demand...with interest annually." On July 1, 1835, Edward Seymour and Austin F. Williams made a $9,000 payment on the note; another payment of $7,596 was made on January 1, 1837, nearly two years after the firm's dissolution. So, although a total of $16,596 was paid on the note, the balance (with interest) had gone unpaid. On March 4, 1838, Timothy Cowles recovered judgment in Hartford County County Court, in the amount of $16,903.30 damages and $8.47 costs of suit, against Edward Seymour, Austin F. Williams, and Timothy Porter, the former partners in the firm of Seymour, Williams & Porter. Dated April 23, 1838 (with a second copy, "No. 2", dated June 30, 1838), the sheriff of Hartford County was commanded to execute the judgment. In copy "No. 2" (physically some 10 feet long), the entire proceeding leading up to the execution was set out.[58]

As of the 1830s Connecticut's judicial process closely resembled that of English common law, to which it owed its origins. It worked like this: if party A believed that party B owed him/her money, A went to the local court and persuaded it to attach sufficient property belonging to B, land or otherwise, to cover the debt. If the property was in the form of goods, the goods were secured pending the outcome of the case. The case then went to court. If A prevailed, the court ordered the sheriff to carry out an execution, in which the sheriff caused the goods to be sold at public auction, after which the proceeds were given to A. If the property was real estate, the execution legally transferred the land from B to A. The execution was then entered on the land records, in the volumes of deeds.[59]

The execution in Case E, *Timothy Cowles vs. Seymour, Williams & Porter*, was repeated in Volume 47, pages 249–255, of the Farmington Land Records. Quoting from the beginning of this record, the following passage was addressed to the sheriff of Hartford County, CT:

...Whereas Timothy Cowles of said Farmington recovered judgment against Edward Seymour, Austin F. Williams, and Timothy Porter, all of said Farmington...for the sum of sixteen thousand, nine hundred and three dollars thirty cents damages, and for the sum of eight dollars forty seven cents costs of suit...whereof execution remains to be done. These are therefore, by authority of the State of Connecticut, to command you, that of the goods, chattels,

Figure 51A. William Bradley, "Successor to Seymour, Hall & Co.," St. Louis, MO, 30-hr. wood movement shelf clock.

Figure 51B. Label of clock shown in Figure 51A.

Figure 51C. Clock shown in Figures 51A and 51B, door open, dial removed. The maker of its movement (Type 8.2, unpublished) is unidentified, although a link to Seymour, Hall & Co. has been discovered. (See text.) (Figures 51A–51C courtesy of Jim and Beverly Lowe.)

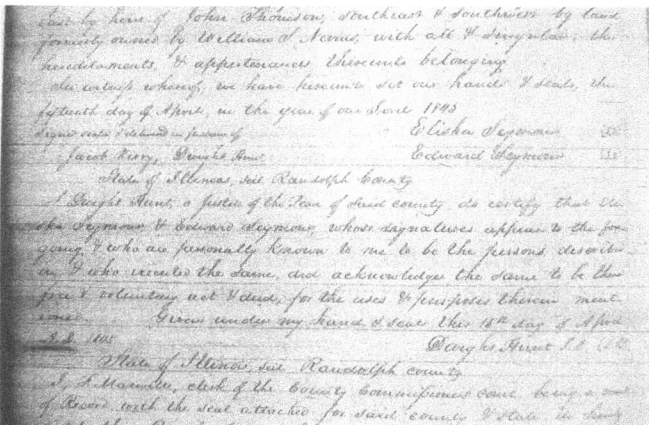

Figure 52A. Signature portion of quitclaim deed dated April 15, 1845, in Farmington Land Records, V. 47, p. 427, identifying Elisha and Edward Seymour as residents of Kaskaskia, Randolph Co., IL.

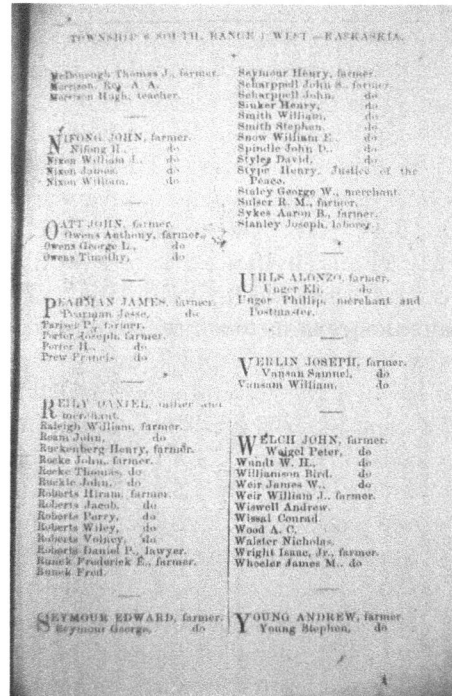

Figure 52B. Page from E. J. Montague's Directory...of Randolph County, Illinois (1859), showing the listing for Edward Seymour, "farmer," and Seymour's sons George and Henry.

or lands of the said debtors, within your precincts, you cause to be levied and the same being disposed of or appraised as the law directs, paid and satisfied unto the said execution the aforesaid sums being sixteen thousand, nine hundred, and eleven dollars seventy seven cents in the whole: with seventeen cents more for this writ... And for want of such goods, chattels, or lands of the said debtors, to be by them shown unto you, or found within your precincts, to the acceptance of the said creditor, for satisfying the aforesaid sums, you are hereby commanded to take the bodies of the said debtors, and them commit unto the keeper of the gaol [jail] in Hartford...within the said prison; who is likewise hereby commanded to receive the said debtors and them safely keep until they pay unto the said creditor, the full sums above mentioned.... Dated at Hartford this 30th day of June Anno Domini, one thousand eight hundred and thirty eight.[60]

Subsequently, the county sheriff assembled his appraisers, whom he dispatched to visit the sites of each piece of property owned by any of the three individual debtors alone, or in combination with any of the others. The appraisers evaluated each piece of property subject to the loss of value caused by mortgages. One of the pieces was the gristmill previously belonging to Solomon Langdon, now owned by Timothy Porter. Four such appraisals were done, and the cumulative real estate value totaled $4,251.17. The amount for execution, originally $16,911.77, as indicated above, with accumulated court charges, was now $17,016.02. The real property, after losses of as much as $40,000 attributable to the ravages of the fire of April 2, was worth only about one-fourth of the amount needed to satisfy the execution, leaving $12,764.85 of the debt unsatisfied.

From the above documents, it appears Timothy Cowles could have had Edward Seymour, Austin F. Williams, and Timothy Porter all sent to prison. However, it was unlikely to have been in Cowles's own interest to do so. In any event, as of August 16, 1838, the date of the third and final execution in the court case, Timothy Cowles owned virtually all of the former Seymour, Williams & Porter property, including all of the former clock-making facilities.

A statement dated February 15, 1840, from Edward Seymour was found in Volume 47 of the Farmington Land Records. It was not a deed, but it reads as follows:

...I Edward Seymour of Farmington...have constituted & appointed and do by these presents constitute and appoint Timothy Porter of said Town of Farmington my true and lawful Attorney for me & in my name, to ask for demand and receive from any person or persons resident in the State of Connecticut or within the United States, all sums of money or other property due or belonging to the old firm of Seymour, Williams and Porter and if necessary in the disposal of any property belonging to the late firm of Seymour, Williams & Porter, to make use of my name in the same manner as if I was personally present and also to use my name in signing a note for my part of a debt or notes due to Major Timothy Cowles from Seymour, Williams & Porter's late Firm; and I do hereby agree to ratify & confirm whatever act or acts my said Attorney may do for the purpose of accomplishing the object of this authority.

In witness whereof I have hereunto set my hand & seal at Tuscaloosa this fifteenth day of Feb. AD 1840.In the presence of
John Hudson
Edward Seymour (L.S.)
Lewis Hudson
State of Alabama Tuscaloosa
County, Tuscaloosa Feb. 15, 1840.[61]

In appointing Timothy Porter as his "Attorney," it appears Edward Seymour was genuinely concerned about the outstanding debts to Timothy Cowles.

As previously mentioned, Edward Seymour assigned his estate before the Farmington District Probate Court for the benefit of his creditors on August 20, 1841. On October 30, Pomroy Strong accepted the position of trustee on the estate, which the Court allowed seven months for settling. Ira Hadsell and Timothy C. Lewis were appointed appraisers.[62] On November 22, 1841, the Court appointed Simeon Hart of Farmington and John Brocklesby of Hartford as commissioners, and six months were allowed for creditors to present their claims. On December 20, 1841, "An Inventory of Edw. Seymour's assigned Estate was exhibited in court, receipts and notes proved and accepted, and placed on file." (But more were added later.) An "Abstract of the Inventory" showed receipts and notes originating in

Tuscaloosa, Sumpter [sic], Jefferson, Bibb, and Fayette Counties [Alabama], discounted down to 10%–25% by the appraisers, totaled to $919.25. On May 20, 1842, the Court allowed an extension of time to April 1, 1843. The "Commissioners' Report" was presented on July 9, 1842 (by S. Hart only, because John Brocklesby had been excused from acting), showing that the total of creditors' approved claims against the estate amounted to $4,659.48. Among the claims were those of "Saxton & Williams," Anson Williams, and J. & A. Cowles. More than half of the total of all claims was the approved claim of Amzi Barnes, in the amount of $2,894.53. The list included no approved claim for Timothy Cowles.

After several delays and time extensions, the Probate Court met on May 4, 1844. Pomroy Strong, the Assignee, presented his account for services of $26.75 (enumerated), for other expenses totaling $21.94, and for the loss on the sale of property (not detailed) of $3,009.93. The total of these was the total negative of the "Trustee's Account," or $3,058.62. On the positive side, details of the $919.25 "Inventory Abstract," above, were given. Additional items were added to the inventory, perhaps more notes and receipts, allowed at 20%, giving $123.83. The last two inventory items listed were "Seymour Hall & Co.," allowed at 20% or $732.04, and "Gillett & Hall," allowed at $12\frac{1}{2}$% or $1,250. These figures gave a total inventory of $3,025.12; to this was added $17.72 for a book account collected. This gave the "Trustee's Account" a total positive balance of $3,042.84, meaning that Pomroy Strong took a loss of $15.78. For the estate of Edward Seymour, the approved claims from the "Commissioners' Report," $4,659.48, were added to Pomroy's loss, giving $4,675.26. Nothing was left on the credit side to distribute to the creditors.[63]

The amount of $732.04 of the inventory of Edward Seymour's estate attributed to Seymour, Hall & Co. may have been for uncollected clock sales by the latter firm. For the firm of Gillett & Hall, the purpose of its presence in the inventory is clear: Seymour, Hall & Co. had been hoping for proceeds in the form of collections from clock sales made by that firm on credit.

On the basis of the information provided above, and in the documents related to Court Cases M and N, Table 2B, for some period during 1839–40, Edward Seymour was himself absent from Connecticut collecting debts in the American South, notably Alabama, due to the firms in which he had been involved. By appointing Timothy Porter as his agent, Seymour set up a mechanism for collecting on outstanding debts

due to Timothy Cowles, and to the other creditors in Seymour's bankruptcy, while he was out of state. After the sheriff's execution of the court's judgment in Timothy Cowles's favor, and the lack of Cowles's name as an approved creditor in Seymour's bankruptcy, it is not clear whether the end of the bankruptcy left Seymour with any legal obligation to Cowles. But some debtors feel moral obligations.

The key to the relationship between Edward Seymour and Timothy Cowles was found late in the second part of the Seymour, Williams & Porter accounts, discussion of which was postponed until now. By 1840 the firm's treasurer, Austin F. Williams, son-in-law of Timothy Cowles, was struggling to close the firm's books. Recalling from Chapter 1 that Richard Cowles was both Timothy Cowles's brother, and a partner in the firm of Cowles, Deming & Camp, on April 1, 1841, the books show, "Note given [to] Richard Cowles on demand & no Claim for principle while he lives but interest to be paid annually 2200.[00]." Then, on June 15, 1843, "Note given up to me [i.e., Williams] in consequence of losses sustained by going into the clock bisness by them [sic] solicitation of T. Cowles - principal 20,000. Int about 10,000–20,000."; and finally on March 1, 1845, "The following Notes given up to me in consequence of losses in the firm of S.W & P. & growing out of it, the business being Entered afore by request of T.C." There follows a list of four notes, dated 1835–37, totaling $25,600.00.[64] Notes "given up" to Austin Williams were the equivalent of canceling them, or in this case, donating $45,600 toward balancing the books of the firm of Seymour, Williams & Porter! We can now look back at each stage of the creation and development of Seymour, Williams & Porter, up to and including its aftermath, and the attempt to settle the firm's accounts, in a new light, as part of Timothy Cowles's enterprises. We can also view Edward Seymour's subsequent efforts to start a new life in frontier Illinois, still peddling clocks as late as the mid-1840s, from the perspective of both his indebtedness to, and sponsorship by, Cowles.

Edward Seymour Starts a New Life

Once the principal partner in his own clock-making firms, Edward Seymour departed from Connecticut, presumably to pursue options other than working in a factory for someone else. As of 1840, inexpensive land in the American West could still be purchased at U.S. government auctions, affording opportunities for many like Seymour, who, for various reasons, determined to leave New England. Since the turn of the century, Randolph County, IL, had been a popular destination for visitors from Farmington, CT, and several former residents settled there. One of them was Edward Seymour's older brother Elisha Seymour (1799–1851), who had made his home in the village of Kaskaskia, Randolph County's picturesque seat, perched on the Mississippi River's east bank. In 1829 Elisha Seymour married Sarah McDonough, the daughter of one of Randolph County's early settlers, at Kaskaskia, where Elisha operated a tannery. Elisha and Sarah became the parents of seven children, all born in Illinois.[65] Hearing of his brother's insolvency, it appears he invited Edward and his family to join him.

According to sources presented in Reference 1 (pp. 302, 304–305), on April 26, 1843, Edward Seymour, "Agent for T Cowles & Cowles & Bradley" (the "Bradley" evidently being William Bradley), left an invoice of property for sale with his brother in Randolph County. The invoice included six Seymour Hall & Co. 30-hr. wood movement clocks in carved cases, four in bronzed and gilt cases, and two 8-day wood clocks produced by the firm (which by that time had been defunct for some five years), "to a/c [account] to him [Edward Seymour] on their order - Elisha Seymour." On October 11 of the following year, Elisha Seymour's wife died, leaving him a widower with a large family.

A deed dated April 15, 1845, recorded on the Town of Farmington Land Records on April 29, described both Elisha and Edward Seymour as "of the County of Randolph & State of Illinois" (Figure 52A).[66] In it brothers Elisha and Edward quitclaimed parcels of land in Farmington, evidently portions of their deceased father's estate, to Farmington relatives. On May 12, 1845, Edward purchased a license to sell clocks for one year in Randolph County. During 1847 Edward Seymour began purchasing land in the state of Illinois.[67]

During this time, while Edward Seymour peddled clocks in the American West, the rest of his family remained behind in Connecticut. Seymour and his family were enumerated in both the 1840 and 1850 U.S. Censuses as Farmington residents. Edward Seymour's occupation as of 1850 was listed in the Census as "joiner."

We have seen that Edward Seymour's business ventures took him to the American South and West for extended periods during the 1840s. But by 1849 the discovery of gold in California was attracting the attention of many Americans with reports of spectacular fortunes made over short periods. Seymour and his older brother were among those who took the chance to improve their lives through speculation on the Pacific coast.[68]

Departing Randolph County in April 1849, brothers Elisha and Edward are said to have journeyed together over land across the plains and over the Rocky Mountains. Reaching California in December 1850, Edward Seymour went to work in the gold mines, with varying results. He and his brother stayed together until May 1851, when Elisha volunteered to ferry a man across the Salmon River at high water. Tragically, although an experienced oarsman and swimmer, Elisha Seymour perished in this undertaking.[69]

San Francisco was still a chaotic shanty town in the summer of 1851, when Edward Seymour took his leave, sailing for New York City reportedly by way of the Isthmus of Panama. Arriving at New York in poor health in 1852, he returned to his former home to recover.[70] Afterward, he removed with his family to Illinois, settling on Elisha Seymour's former estate. Edward Seymour was residing in Kaskaskia when, on July 17, 1854, in response to his request for legal advice, a prominent young Springfield, IL, lawyer and politician named Abraham Lincoln wrote to Seymour, recommending the latter send him evidence of title to a parcel of land. Whether Seymour pursued the matter is unknown.[71] We will encounter Lincoln again in Chapter 5.

By 1859 Edward Seymour was engaged in agricultural pursuits in Kaskaskia, as were his sons George (1833–1913), and Henry (1833–1907) (Figure 52B).[72] He continued to accumulate land in Illinois, eventually owning a total of over 300 acres in Randolph and Washington counties.[73]

At the time of the 1860 U.S. Census, Edward Seymour was still farming and living with his wife, Harriet, and sons Henry and George D. Seymour, at Kaskaskia. He appeared in the 1865 Illinois State Census as a Kaskaskia resident. Mrs. Harriet (Johnson) Seymour died in 1872, after an illness that lasted about six months. On July 23, 1879, Edward Seymour died of heart failure at Ellis Grove, IL. He was 78 years old.[74]

Reference 1 (pp. 300–305) outlined a portion of Edward Seymour's clock-making activities. The present authors have uncovered and analyzed many new sources, including numerous court records, tax records, records of Edward Seymour's bankruptcy, and of his move to Illinois. In addition to uncovering the names of the partners in Seymour, Hall & Co., the present work corrects some errors and adds considerable depth and dimension to the story thus far.

In Chapter 4, we will return to Unionville village to examine what happened after the disastrous end of Edward Seymour's firms. Remarkably, despite a distressed national economy and discouraging losses, Timothy Cowles and others remained optimistic about the future of clock making at Unionville.

Notes and References

1. Kenneth D. Roberts and Snowden Taylor. *Eli Terry and the Connecticut shelf clock.* 2nd ed. Fitzwilliam, NH: Ken Roberts Publishing Co., 1994.

2. Farmington Land Records (FLR), V. 47, pp. 154–155, August 7, 1835.

3. FLR, V. 46, p. 305, September 5, 1835.

4. Calvin Cowles (compiler). *The Cowles families in America.* New Haven, CT: Tuttle, Morehouse & Taylor, 1929.

5. Mabel Hurlburt. *Farmington town clerks and their times.* Hartford, CT: Finlay Press, 1943.

6. Julius Gay (compiler). "Maps of Old Farmington." Unpublished photocopy. Connecticut State Library.

7. *Connecticut Courant,* July 27, 1835.

8. FLR, V. 47, pp. 151–152, September 8, 1835.

9. FLR, V. 46, p. 303, September 8, 1835.

10. FLR, V. 46, p. 441, September 10, 1836.

11. FLR, V. 46, p. 573, February 15, 1839.

12. Julius Gay (compiler). "Maps of Old Farmington."

13. Mabel Hurlburt, *Farmington town clerks and their times,* 1943.

14. FLR, V. 45, p. 30, May 11, 1830.

15. FLR, V. 47, pp. 151–152, September 8, 1835.

16. FLR, V. 47, pp. 214–215, September 12, 1835.

17. FLR, V. 46, p. 323, November 30, 1835.

18. Records of the Town of Farmington, Individual Tax Returns, Box 12, RG62, Connecticut State Library.

19. Ibid. It may be useful to note that a usual assessment date in this part of Connecticut at about this time was October 1 of a given year.

20. Ibid. See also Box 11.

21. *Joseph R. Gillett vs. Frederick P. Hall, Edward Seymour, Edward Sexton, and Anson Williams.* Litchfield County Superior Court Files, Box 219, RG3, Connecticut State Library.

22. Ibid.

23. Ibid.

24. Ibid.

25. Ibid.

26. Ibid. According to the court case documents, the partners in Seymour, Hall & Co. were represented by attorneys Ellsworth and Woodruff.

27. Ibid.

28. Ibid. The fact that in pursuit of its business, the firm of Gillett & Hall (or by extension, Seymour, Hall & Co.) owned a slave, in apparent conflict with the personal views of several Farmington and Unionville residents closely connected with the clock-making enterprise, provides an example of the moral conflicts facing many Northerners doing business in the antebellum American South.

29. Thomas Day. *Reports of cases argued and determined in the [CT] Supreme Court of Errors.* (Originally published 1841; reprinted 1960): 427–436.

30. Donald Jacobus. *A history of the Seymour family.* New Haven, CT: Tuttle, Morehouse & Taylor, 1939.

31. *Harlow Gillet vs. Gillet & Hall.* Litchfield County County Court, Papers by Subject, Executions, RG3, Connecticut State Library.

32. Ibid.

33. Ibid.

34. *Harlow Gillet vs. Gillet & Hall.* Litchfield County County Court Records, V. 21, p. 345, Connecticut State Library.

35. *Harlow Gillet vs. Gillet & Hall.* Litchfield County County Court, Papers by Subject, Executions, RG3, Connecticut State Library.

36. *Joseph Gillet vs. Gillet & Hall.* Town of Farmington, Justice of the Peace Records, Box 41, RG62, Connecticut State Library.

37. Ibid.

38. *Darrow vs. Seymour, Hall & Co.* Hartford County County Court Files, Box 457, RG3, Connecticut State Library.

39. See, for example, FLR, V. 45, p. 453 and V. 47, p. 427.

40. *Connecticut Courant*, November 24, 1834.

41. Dorothy Dunne. *Industrial history of Unionville.* Unpublished Master's thesis. Central Connecticut State University, 1968.

42. *Waterbury American*, February 16, 1855.

43. *Connecticut Courant*, May 2, 1838.

44. 1840 and 1850 U.S. Censuses, Farmington, Connecticut.

45. Snowden Taylor identified Seymour, Williams & Porter as a movement maker in his "Update on 30-hr. Wood Terry-type Movements." NAWCC *Bulletin*, No. 237 (August 1985). A later (unpublished) update designates Seymour, Hall & Co. as the maker of a similar movement, Type 1.412, in which the count wheel pinion bridge is constructed and arranged like the escape wheel bridge.

46. Ibid.

47. Bryan Rogers and Snowden Taylor. "Eight day wood movement shelf clocks: their cases, their movements, their makers." NAWCC *Bulletin* Supplement No. 19 (Spring 1993).

48. Snowden Taylor, in his "Update on Terry-type 30-hour wood movements." NAWCC *Bulletin* No. 249 (August 1987): 314–315, also mentions Seymour, Hall & Co. as the user of a movement designated Type 5.113, made by Marsh, Gilbert & Co. or George Marsh & Co. Both of the latter were firms of Farmington village until 1834. It is suggested that Seymour, Hall & Co. may have obtained a few of these movements from Williams, Orton, Prestons & Co., a successor to the Marsh firms. (See Court Case L, Table 2B.) See Chapter 1 for further information on the Marsh firms and their successors.

49. Calvin Cowles (compiler). *The Cowles families in America,* 1929.

50. FLR, V. 47, p. 229, October 9, 1837.

51. *Chauncey D. Cowles vs. Seymour, Hall & Co.* Hartford County County Court Files, Box 436, RG3, Connecticut State Library.

52. According to William Bridgewater and Elizabeth Sherwood, eds. In: *The Columbia Encyclopedia.* 2nd ed. NY: Columbia University Press, 1950, Horace Greeley (1811–72) and a partner founded the weekly journal *The New Yorker* in 1834.

53. The authors are grateful to Clifford Alderman of the Unionville Historical Society for locating the notices of the fire in both the *Connecticut Courant* and *The New Yorker*.

54. FLR, V. 45, p. 256, May 18, 1838.

55. Plymouth Vital Records, V. 1, p. 21.

56. Bristol, CT First Congregational Church Records, V. 2, p. 219; New Britain Cemetery Records; and 1840 U.S. Census, Farmington, CT.

57. FLR, V. 46, p. 573, September 10, 1836; and 47, p. 290, December 28, 1839.

58. *Timothy Cowles vs. Seymour, Williams, & Porter.* Hartford County County Court, Executions, Box 529, RG3, Connecticut State Library.

59. Zephaniah Swift. *A digest of the laws of the state of Connecticut.* Vol. 2. (New Haven, CT: S. Converse, 1822).

60. FLR, V. 47, pp. 249–255, June 30, 1838–August 16, 1838.

61. FLR, V. 47, p. 311, February 15, 1840.

62. Based on information found in the Seth Wheeler Papers (RG69:47, Connecticut State Library), this is likely the same Timothy C. Lewis of Farmington who peddled C. & L. C. Ives' clocks in Kentucky between 1836 and 1840, while employed by Seth Wheeler & Co. At that time, Lewis was regarded as one of the firm's best peddlers.

63. Farmington Probate Records, V. 10, pp. 263–264; 271–273; 286; 292; 295; 325; 386–387; and "Edward Seymour's Assigned Estate," Farmington Probate District Estate Papers, Connecticut State Library; hereinafter referred to as "Probate Records of Edward Seymour."

64. "Seymour, Williams & Porter Account Book 1831–1850 Unionville Clocks." (Modern designation.) Vols. 1 and 2, MS91865, Connecticut Historical Society.

65. George Dudley Seymour, compiler. *A history of the Seymour family.* (New Haven, CT, 1939).

66. FLR, V. 47, p. 427, recorded April 29, 1845.

67. State of Illinois, Public Land Purchase Records 1813–1909. Ancestry.com (online source).

68. W. R. Brink & Co. *An illustrated historical atlas map of Randolph County, Illinois.* Philadelphia, PA: W. R. Brink & Co., 1875.

69. Ibid.

70. *Portrait of Randolph, Jackson, Perry, and Monroe Counties, Illinois, containing biographical sketches.* Chicago, IL: Biographical Publishing, 1894: 429.

71. As of 2009, the letter was in the collection of a Mrs. Arthur Seymour of Fort Gage, IL, likely a descendant. See also: *Reports of cases determined in the Supreme Court of the State of Illinois, January 1859.* Chicago, IL: D. B. Cook & Co., 1860.

72. E. J. Montague. *A directory, business mirror, and historical sketches of Randolph County.* Alton, IL: Courier Steam Book and Job Printing House, 1859.

73. Ibid.

74. Randolph County Death Register, Illinois Regional Archives Depository, Carbondale, IL.

Chapter 4

The Unionville Firm of Williams, Orton, Prestons & Co.

Contributing much new information to the framework proposed in Reference 1,[1] Chapter 1 of the present work told the stories of the related firms of Marsh, Gilbert & Co., followed by George Marsh & Co. (wherein the "& Co." was Cornelius R. Williams), followed in turn by Orton, Preston & Co. (partners now known; Williams not among them). Orton, Preston & Co. probably began operating in April 1834, in Cowles, Deming & Camp's former textile mill in Farmington village. As explained in Chapter 1, the successor to all of these firms was Williams, Orton, Prestons & Co., the "Williams" once again being Cornelius R. Williams. Another change was the addition of a second member of the Preston family: Eli Dewey Preston (1804–87), Noah Preston's younger brother. Following the death of his first wife, Phebe (Merwin), in Harwinton on February 28, 1833, Eli D. Preston married (2nd) Chloe Mallory of Barkhamsted in 1834,[2] moving to Farmington shortly afterward to join the firm.[3] As in the case of the three earlier firms, no deeds have been found for Williams, Orton, Prestons & Co. during the remainder of the 1830s, so it must have rented manufacturing space.

Several key questions remain to be answered: When did the firm of Williams, Orton, Prestons & Co. commence? Was there a smooth transition from the ending date for Orton, Preston & Co.? Did the new firm start within Cowles, Deming & Camp's mill in Farmington village? Did the firm move from that mill to Unionville village before the end of the 1830s, and if so, where was it located? What sorts of clocks did the firm make? Who were the people involved, and what were they like? What was the ending date for Williams, Orton, Prestons & Co.?

Some information relative to the above questions was summarized in Chapter 1 near the end of the discussion of the firm of Orton, Preston & Co. A "weak" date for the ending of Orton, Preston & Co. was obtained, suggesting "not later than August 15, 1835." As explained in Chapter 1, data from an earlier study of these problems

suggested "about 1836 with a smooth transition" for the ending of Orton, Preston & Co. to the beginning of Williams, Orton, Prestons & Co., and a firm date of "not later than 1837" for the beginning of Williams, Orton, Prestons & Co. The authors also established a tentative link between the firm of Orton, Preston & Co. and Cowles, Deming & Camp's mill in Farmington village. According to local historian Mabel S. Hurlburt (1943):

> 1820, [sic] Williams, Orton Preston organize in Farmington to manufacture clocks. (This was in the brick building back of present drug store...) Afterwards move the business to Unionville and occupy the old screw shop of Pierpont & Tolles. Clock business abandoned about 1837.[4]

Examining this statement, the authors agreed that Williams, Orton, Prestons & Co. (or at least its predecessor firm, Orton, Preston & Co.), probably started at the building indicated, part of Cowles, Deming & Camp mill, discussed more fully in Chapter 1. However, Hurlburt was apparently unaware that several successive clock businesses operated at that location; the first was Marsh Gilbert & Co, beginning in January 1832, not 1820. Furthermore, because no leases were found to establish that Williams, Orton, Prestons & Co. ever occupied the old Patent Wood Screw Manufactory of Pierpont and others, the authors were initially skeptical about the possibility. In addition (as we will see), Williams, Orton, Prestons & Co. did not give up the business "about 1837" as Hurlburt indicated.

Searching for answers, the authors examined numerous sources, including the records of several court cases discovered by author MJD, in which the firm of Williams, Orton, Prestons & Co. was involved. These are summarized in Table 6A (an extension of Tables 2A and 2B, in Chapters 2 and 3, respectively). We refer to Table 6A often during our discussion of Williams, Orton, Prestons & Co.

Table 6A. Court cases involving the firm of Williams, Orton, Prestons & Co.

The date of the writ and summons initiating the court case(s) is provided under "Date of Writ(s)". If the dispute was over non-payment or partial non-payment of a note, the amount of the note (if known) is given under "Amt. of Note"; the date of the note is listed in parentheses below. A summary of the outcome of the cases, as far as known, is provided as "Outcome / Additional Info." Unless otherwise noted, the outcome was in favor of the plaintiff. Full references for each case are provided under "Reference." "Exn." = the date on which the court granted execution of its judgment (not necessarily the same as the date of the actual execution.) "Plff" = plaintiff; "Deft" = defendant; "Default" means the defendants failed to appear, in which case the court customarily rendered judgment in favor of the plaintiff; "SH & Co." = Seymour, Hall & Co.; "WOP & Co." = Williams, Orton, Prestons & Co.; "Htfd. Co." = Hartford County; "Ltfd. Co." = Litchfield County; "Sup." = Superior; "CSL" = Connecticut State Library; "CHS" = Connecticut Historical Society. All amounts are in dollars.

Case Date of Writ(s)	Title of Case	Amt. of Note / (Date of Note)	Outcome / (Additional Info)	Reference
		Group IV: Williams, Orton, Prestons & Co. Cases		
AA 9/16/1836	Samuel E. Judd v. WOP & Co.	[Book debt] (Unknown.)	Withdrawn 3/1837.	Htfd. Co. Co. Court Files, Box 430, RG3, CSL.
BB 3/12/1839	WOP & Co. assigned to Wm. Beach v. SH & Co.	100.00 (2/1/1838)	Default; Exn.4/5/1839; (Note assigned to Beach 8/7/1838.) (Same as Case L, Table 2B.)	Htfd. Co. Co. Court Files, Box 437, RG3, CSL.
CC 8/29/1840	WOP & Co. v. Daniel Tuttle	? 114.00 (4/13/1840)	Default; Exn. 3/25/1841; deft out of state; may request new trial. (Heman Orton, agent for deft.)	Htfd. Co. Co. Court Files, Box 441, RG3, CSL.
DD 2/8/1841	WOP & Co. v. John Hunt	98.40	Default; Exn. 3/25/1841	Htfd. Co. Co. Court Files, Box 442, RG3, CSL.
EE 5/1/1841	Ira Bowen v. WOP & Co.	74.96 (12/7/1837) 74.96 (1/7/1839) 49.21 (3/2/1840)	Default; Exn. 11/14/1841.	Htfd. Co. Co. Court Files, Box 443, RG3, CSL; Farmington Justice of Peace Records, Box 39c, RG62, CSL.
FF 11/30/1841	WOP & Co. v. Daniel F. & George M. Olcott	438.00 (11/18/1840)	Judgment in favor of defts to recover their court costs of 1.79.	Farmington Just. of Peace Records, Box 41, RG62, CSL.
GG 3/16/1842	WOP & Co. v. John O. Pettibone & Aurora Case	500.00 (Book debt)	Withdrawn 9/1842. (WOP & Co. copartners".)	Htfd. Co. Sup. Court Files, Box 96, RG3, CSL.
HH 6/22/1842	WOP & Co. (except Noah Preston) v. Noah Preston	[See text.]	Default; withdrawn 9/1843. (WOP & Co. "continued in business until about" 1/1/1842.)	Htfd. Co. Sup. Court Files Box 99, RG3, CSL; and Farmington Just. of Peace Records, Box 40, RG62, CSL.
II [Unknown]	WOP & Co. v. Oliver Weldon	36.00? (Unknown.)	Exn. 11/10/1842. (WOP & Co. "late copartners". See text.)	Htfd. Co. Co. Court Exns., Box 533, RG3, CSL; Htfd. Co. Co. Court Files, Box 444, RG3, CSL.
JJ 8/10/1843	WOP & Co. v. Wm. Tyler	216.00 (10/22/1839) 288.00 (11/5/1840)	Default; Exn. 3/1844. (2nd note given to pay 1st one; WOP & Co. "late copartners".)	Htfd. Co. Co. Court Files, Box 446, RG3, CSL.
KK 11/25/1843	WOP & Co. v. William Daniels	250.00 (Book debt.)	Withdrawn 8/1844. Deft. to recover court costs.	Htfd. Co. Co. Court Files, Box 447, RG3, CSL. (WOP & Co. "late manufacturers & dealers...")

By far the most important of the court cases in Table 6A in answering the outstanding questions about the firm was Case HH, in which the all of the partners in Williams, Orton, Prestons & Co. as of June 22, 1842, except Noah Preston, brought suit against the latter. The matter of why the suit was brought is taken up below, at an appropriate point chronologically. For present purposes, however, the records of this case yielded the following previously unknown account of the firm's history:

> The Petition of Cornelius R. Williams, Heman Orton & Eli D. Preston of Farmington...and Henry C. Bull of Alton... Illinois humbly Sheweth that your Petitioners together with Noah Preston, Benham Beecher & Frederick W. Crumm [sic] of said Farmington on or about the 15th day of December 1835, formed a Connection in business to wit of Manufacturing and Vending clock[s] to be conducted and Carried on in said Farmington for the term of three years from s^d Date last aforesaid under the name of Williams, Orton, Prestons & Co and after deducting the expenses of said business the profits and losses were to be Shared between the respected [sic, respective] members of said Company in proportion to their interests therein that the aforesaid interests in said Company was divided into thirteen Shares of which the said Williams became the proprietor of four Shares the s^d Heman, Noah, & Eli D. two shares each and the [said] Henry C., Benham & Frederick W. of one Share each, that Subsequent to the original arrangement so entered into as aforesaid the said Benham and Frederick W. released and transferred all their interest in s^d Company to the Petitioners who together with said Noah Preston under and by virtue of the Connexion so entered into as aforesaid engaged extensively in the manufacture & Sale of Clocks...[5]

Thus, the date on which the firm of Williams, Orton, Prestons & Co. commenced, given in this account as "on or about" December 15, 1835, is in good agreement with the "weak" ending date for Orton, Preston & Co. of August 15, 1835, and not far from the "about 1836" given in Chapter 1. It can now be said with confidence that the transition between the two firms occurred in "late 1835" and was likely smooth. The court case confirmed the principal partners in Williams, Orton Prestons & Co., but in addition, contributed the important discovery of the identities of several new individuals who comprised the "& Co.," including Henry C. Bull, a resident of Alton, IL, as of June 22, 1842, when the suit was initiated; Benham Beecher (a holdover from Orton, Preston & Co.); and Frederick W. Crum.

In addition to providing evidence of the firm's starting date, although Frederick W. Crum was no longer a partner in Williams, Orton, Prestons & Co. as of the commencement of Court Case HH, its discovery was particularly important because it revealed that he had in fact once been a partner in the firm. Moreover, Crum was still named as a partner on September 16, 1836, when Samuel E. Judd of Farmington brought suit against the firm (Case AA, Table 6A), although Benham Beecher was not. In his suit Judd complained that Williams, Orton Prestons & Co. (with partners identified as Cornelius Williams, Heman Orton, Noah Preston, Eli Dewey Preston, Frederick W. Crum, and Henry C. Bull, "all of said Farmington") had failed to pay $80 the firm owed him on his book accounts. On September 20, 1836, Farmington constable Pomroy Strong attached as the firm's property "one hundred unfinished Clock Cases, & also the movements of one hundred thirty hour Clocks put together..."[6] However, the nature of the debt was unspecified, and the suit was later withdrawn.

The individual partners in Williams, Orton, Prestons and Co. were not named in Court Case BB, and Frederick W. Crum was not listed among the firm's partners who initiated Court Case CC on August 29, 1840. Therefore, Crum probably left the firm sometime after September 16, 1836, but before August 29, 1840. All this is interesting in view of the outstanding questions about Williams, Orton, Prestons & Co., because on July 3, 1890, some 50 years later (and five years prior to his death), Frederick W. Crum's recollections of the clock industry in Unionville were published as part of an article in the *Bristol* [Connecticut] *Herald* newspaper. The relevant portion of the article is quoted as follows:

> ...F. W. Crum and Mr. Barbour formed a partnership about 1835. They bought the [Patent Wood] Screw company's property, also that of Marsh & Gilbert, clock makers in Farmington and moved the business from Farmington to the Screw Shop. A company by the firm name of Williams, Orton & Preston was making eight-day weight brass clocks in

1836 and Crum & Barbour bought a portion of the business in 1841, Mr. [Heman] Orton purchasing the case department of that company and made cases for the movements manufactured by Crum and Barbour... The firm of Seymour, Williams & Porter was burned out in 1837 [sic; the fire actually occurred on April 2, 1838], but the firm of Crum & Barbour continued to make clocks here [Unionville] until 1844, when the price of clocks, which had formerly been $18 got so low that they went out of the business...[7]

We will take up Frederick W. Crum's, and Crum & Barber's, clock-making activities later, in Chapter 7. However, for present purposes, Crum's remarks seemed to provide a firsthand report that the firms of Crum & Barber, and perhaps Williams, Orton, Prestons & Co. (in which Crum is now known to have been an early partner), indeed operated within the former Patent Wood Screw Manufactory in Unionville during the mid- to late-1830s.

The firm of Williams, Orton, Prestons & Co. did not own any land when it appeared on the town of Farmington's tax lists of 1837, 1838, 1839, and 1840 and was assessed total tax bills of $308.30, $215.00, $175.00, and $118.00, respectively,[8] so it must have been renting manufacturing space. The data do not necessarily reflect declining fortunes.[9] In contrast, the firm of Cowles, Deming & Camp, in whose Farmington village mill Williams, Orton, Prestons & Co.'s predecessor firm, Orton, Preston & Co., likely began, was assessed $133.50 in 1837, and at the same level in 1839. In each of these years Cowles, Deming & Camp's assessed properties included "1 Manufactories," valued at $800, but the 1837 assessment noted "shop not occupied."[10] If Farmington's assessments were made as of October 1 (as they were for many other Connecticut towns during the 1830s), then the statement "shop not occupied" written in Cowles, Deming & Camp's 1837 assessment meant that no manufacturing had taken place there during the preceding year (i.e., October 1836–October 1837) or at least that the shop was unoccupied as of October 1837. So if Williams, Orton, Prestons & Co. had started out renting Cowles, Deming & Camp's manufactory, they had relocated by that date.

On October 16, 1837, John T. Norton, Abner Bidwell, and J. & A. Cowles, then co-owners of the former Patent Wood Screw Manufactory, transferred a small parcel of land in close proximity thereof in Unionville, to John S. Preston, a brother of Eli D. and Noah Preston. The price was $20.[11] This transaction is interesting for two reasons, the first being that, in his letter to a clock collector dated February 25, 1927, Edward N. Preston of Unionville, a grandson of Eli D. Preston, stated that "John Preston" had once been a partner in Williams, Orton, Prestons & Co.[12] A family genealogy also stated that John S. Preston "worked in the clock business" in Unionville for a time,[13] although the authors have seen only circumstantial evidence connecting him with the firm. On April 23, 1838, Preston sold the parcel, now with buildings standing thereon (presumably residential in nature), to Benjamin C. Hosford of Avon for $300,[14] returning to the Prestons' hometown of Harwinton by 1840.[15] Therefore, it appears that John S. Preston was indeed involved with Williams, Orton, Prestons & Co., if only for a relatively short time. The second reason is the location of the parcel in question: in 1835, following a dispute over the availability of waterpower, Sherman Pierpont and his partners abandoned the manufacture of wood screws in the nearby factory building, newly built for that purpose in 1832.[16] Again, no deeds were found to document subsequent leases of the building. However, in addition to offering to lease manufacturing space within the building, an advertisement placed by Norton, the Cowles brothers, and Bidwell, in the *Connecticut Courant* of July 30, 1835, offered a total of five dwelling houses "conveniently situated for persons renting the factory Building" for sale or rent (Figure 53). Therefore, it seemed significant that one of the Preston brothers had purchased land nearby in October 1847.

A letter from a firm by the name of Christie & Chamberlain, Deckertown [New Jersey?], dated May 17, 1839, addressed to "Williams, Orton, Prestons & Co., Unionville," provided the first real documentation encountered by the authors for the firm's location in the village. The letter reads:

Gent. your of the 4th May [1839] is now before us informing us that you have sent those four Boxes clocks to Carbondale [Illinois] which is all right. And are sorry to say we have been unable to send you money until now. We enclose for you one fifty Dollar bill on the Essex Bank Co.[?] New Mesa[?] which you will please place to our credit. Business is tolerable fair_ We want you to send two Boxes more half glasses & half landscapes to use [sic] Care of Oakly & Davis Newburgh.

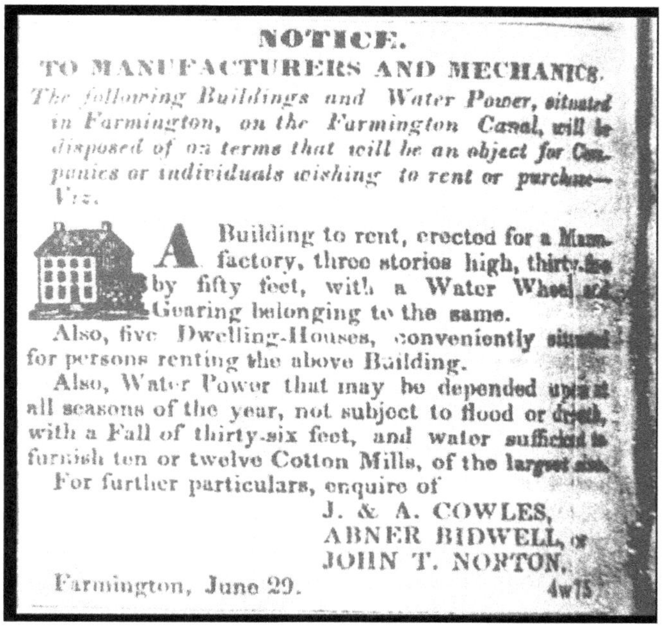

NOTICE.
TO MANUFACTURERS AND MECHANICS.

The following Buildings and Water Power, situated in Farmington, on the Farmington Canal, will be disposed of on terms that will be an object for Companies or individuals wishing to rent or purchase— Viz:

A Building to rent, erected for a Manufactory, three stories high, thirty-five by fifty feet, with a Water Wheel and Gearing belonging to the same.

Also, five Dwelling-Houses, conveniently situated for persons renting the above Building.

Also, Water Power that may be depended upon all seasons of the year, not subject to flood or drouth, with a Fall of thirty-six feet, and water sufficient to furnish ten or twelve Cotton Mills, of the largest size.

For further particulars, enquire of
J. & A. COWLES,
ABNER BIDWELL, or
JOHN T. NORTON.
Farmington, June 29.   4w15

Figure 53. Advertisement offering for rent the facilities formerly occupied by the Patent Wood Screw Manufactory, Connecticut Courant, July 20, 1835.

Gent We shall keep up our cash payments as well as possible. Money is very remarkably scarce in our vicinity.

We dare not promise money on the clocks you send to Carbondale under one year for fear of disappointing which is or would be much worse than to tell you the Honest truth. Gent. If you do trust [us] as to clocks what we want to sell at Carbondale you are certainly safe for your money and we think it perfect folly for us to promise money unless we think we can fulfil...[17]

On the basis of the information provided in Christie & Chamberlain's letter, the nation's economy was still suffering from the effects of the severe depression precipitated by the Panic of 1837, complicated by the lack of a national currency. Despite the need for long credit, however, at least a few customers were still ordering clocks!

A pair of letters found in the American Clock & Watch Museum's collection, both addressed to "Williams Orton Preston & Co.," also spoke of clock purchases, past or planned.[18] The first, dated June 30, 1839, made no mention of the firm's location. It was signed by "H.P. Horton" and sent from "South Middleton" [probably Pennsylvania]. However, the second letter, dated December 19, 1839, signed by Stephen Colegrove of Hartford, almost certainly the Stephen Colegrove (or Colegrave) who was at one time the partner of clock casemaker John Hunt of Plainville village,[19] was also addressed to "Williams Orton Preston & Co." at Unionville. So, considered cumulatively, the data suggested Williams, Orton, Prestons & Co. moved to Unionville sometime prior to May 17, 1839, the date of Christie & Chamberlain's letter, but perhaps as early as 1836 or 1837, based on the tax assessment data for Cowles, Deming, & Camp's then unoccupied mill.

Searching further, a historical account of Unionville written in 1891[20] noted that at one time (i.e., approximately 1836–37), there were *two* clock manufactories in the village. The larger one, described as "standing on the present site of the Sunny Side mill [i.e., in 1891]," was said to have employed about 30 men. If true, this account would seem to place Williams, Orton, Prestons & Co. as operating within the former Patent Wood Screw Manufactory building, a portion of which was occupied as of 1891 by the Sunnyside Paper Mill. However, the oldest portion of that mill, containing the original screw factory (on present-day Perry Street) and afterward used by the C. W. House Co. for felt making, was destroyed by fire in 1908.[21]

The same historical account went on to describe the smaller of the two clock-making firms as employing about 20 men, and "...having its two buildings on the present site of the Sanford & Hawley factory & the Keys shop for the manufacturing of cases and movements, respectively." Beginning about 1848, David A. Keys occupied a building on the site of Seymour, Williams & Porter's former western clock factory building, which he used to manufacture cutlery.[22] So the clock-making facilities of Edward Seymour's firms, of which the [western] movement shop had been destroyed in the fire of April 2, 1838, undoubtedly comprised the smaller of the two clock manufactories described in the account.

Definitive answers to the questions of when Williams, Orton, Prestons & Co. moved to Unionville, and whether the firm had indeed occupied the former screw factory, were finally located in extant Town of Farmington tax records. According to these records, by 1835 the firm of Norton, Cowles & Bidwell[23] was the owner of one factory building in "Union[ville]," undoubtedly the former screw factory, plus two nearby dwelling houses (apparently out of the original five) reserved for rental by the factory's workers. In 1835 one of the two dwelling houses was still occupied by "More," perhaps referring to Oliver Moore, who had until recently been one of Sherman Pierpont's partners in the Patent Wood Screw Manufacturing Co.[24] The other house was "occupied by Preston[,] unoccupied the rest of the year."[25] So one of the Preston brothers had moved into the house during the latter part of 1835, about the time the transition from Orton, Preston & Co. to Williams, Orton, Prestons & Co. occurred.

According to Norton, Cowles & Bidwell's 1836 tax record, the firm "Prestons & Co." (undoubtedly, Williams, Orton, Prestons & Co.) then occupied the two dwelling houses "near the factory." Beginning in 1837 the records designated the latter building as "the clock factory." The building continued to be referred to as "the clock factory" in the town tax records through 1847, about five years after the firm of Williams, Orton, Prestons & Co. dissolved. In 1848 the building's description reverted to "the old screw factory." The significance of the year 1847 in the history of clock making in Unionville village will be made clearer shortly.

"Prestons & Co." continued to occupy the two dwelling houses that remained associated with the former screw factory through 1840. In 1841 one of the houses near the clock factory began to be "occupied by Bancroft," and the second house was occupied by "Russell." Bancroft, identified in later records as "Henry Bancroft," continued to occupy one of the rental houses through the year 1847. (More on Bancroft and his connection with Williams, Orton, Prestons & Co. later.) The second house, simply described as "occupied" in 1842, was subsequently rented by a "Mr. Stone"[26] as of the 1843 assessment and simply "occupied" thereafter, until at least 1848.

Following these discoveries, information was found that not only shed light on the character of some of Unionville's clockmakers but also connected them with the antislavery views shared by a few prominent members of the local community.[27] For example, during June 1838, Noah Preston, Henry C. Bull, and Heman H. Orton, three of the partners in Williams, Orton, Prestons & Co., together with John S. Preston and others, were among those who entered subscriptions to the Connecticut Antislavery Society's fledgling newspaper, the *Charter Oak*.[28] Furthermore, except for John S. Preston, their names, and the names of several additional known clockmakers, including Frederick [W.] Crum, Levi Smith, and Isaac P. Frisbie, appeared on one

or more of a series of four petitions to Connecticut's General Assembly made by Farmington residents during 1838 in support of what were then highly controversial measures. These measures aimed, for example, at eliminating "distinction among...[the State's] inhabitants on account of color," prohibiting admission of new slave states to the Union, and ensuring trial by jury for fugitive slaves. At least two of the clockmakers, Henry C. Bull and Heman H. Orton, signed all four of the petitions.[29] Although the State legislature failed to act on the petitions, the records attest to the men's remarkable courage and foresight in these matters.

Court Case BB, initiated on March 12, 1839, linked the clock-making firm of Seymour, Hall & Co. with Williams, Orton, Prestons & Co., both of Unionville, and also with William Beach, a clock dealer of Farmington and Hartford. The subject of the suit was a note Seymour, Hall & Co. had given to Williams, Orton, Prestons & Co. on February 1, 1838, in payment for unspecified goods or services, perhaps including clock movements or parts. The latter firm assigned the note to Beach. This case was discussed in Chapter 3, where it appeared as Court Case L in Table 2B.

Meanwhile, a complex arrangement was falling into place involving the former clock-making facilities of Edward Seymour's firms on Roaring Brook at Unionville. Recalling from Chapter 3 that during October 1837, Seymour, Hall & Co.'s stock and tools, and the land and buildings it occupied, were placed under attachment in a number of court cases, culminating with Seymour and Williams's sublease of the facilities and lands to C. D. Cowles on October 9. Subsequently, on April 2, 1838, the western factory building (i.e., the movement shop of Edward Seymour's firms) was severely damaged by fire, and by August of the same year Timothy Cowles acquired the entire property for unpaid debts as part of the execution of a court's judgment in his favor against Seymour and Williams.[30]

The new arrangement was revealed by means of two very odd deeds, both dated at Farmington, December 28, 1839. The first reads:

> In consideration of one dollar received to my full satisfaction of Timothy Cowles, I do hereby lease unto him, and by this act, give unto him possession of a certain lot of land and the buildings thereon, now occupied by me at Unionville, and which I lately purchased of him the said Cowles

for which the said Cowles is to pay me a reasonable rent, and is to have and hold the said premises, until the first day of October AD. 1840.[31]

This deed was signed by William Bradley. As mentioned previously, Bradley was not only Edward Seymour's brother-in-law[32] and "agent" but also Seymour, Hall & Co.'s "successor." So it appears Bradley had been wrapping up unfinished clock business at Roaring Brook, in connection with Timothy Cowles.[33]

The second deed, written with the grantees in the plural rather than in the singular as above, was identical up through the words "possession of" and then continued: "the factory building which we now occupy in the prosecution of our manufacturing in Unionville, and which we hold under a lease from the said Cowles:"[34] It was signed "Markland, Bradley & Co." Evidently, the "Bradley" in the firm was William Bradley. The "Markland" in Markland & Bradley was probably William Markland, Jr. (1814–41), who arrived in the United States from Liverpool, England, on July 21, 1827, with his father, William Markland Sr. (1788–?), a glass and pottery dealer,[35] his mother, Susan (ca. 1792–?), and several siblings.[36] At a meeting of Farmington's "Selectmen and Board of Civil Authority" on April 2, 1838, William Jr. and William Sr., "naturalized citizens of the U.S.," were admitted as inhabitants of the town.[37] Markland & Bradley had been manufacturing silver-plated ware in Seymour and Williams's former eastern clock factory building. On July 19, 1841, their tools and equipment were acquired by Edward K. Hamilton, who had also been manufacturing hooks and eyes in the building.[38] In both of the aforementioned deeds, the occupants, with leases (not found in the deeds), leased the property back to Cowles, its owner. It would therefore appear that Cowles was repossessing part of his factory complex with some purpose in mind, perhaps having to do with the planned sale, described below.

On January 13, 1840, Timothy Cowles deeded to Cornelius R. Williams, Noah Preston, and Eli D. Preston (Heman Orton was not named; apparently, he was not a party to the firm's Connecticut land transactions) a portion of the property he had obtained by execution from the former partners of Seymour, Williams & Porter. The purchase price was $3,300. The parcel was described as:

> ...one certain piece of land, at Unionville in said town [Farmington], containing about twenty acres, be the same more or less, with two dwelling houses thereon, bounded as follows, viz: North on land of George W. Payne in part, and in part of land of Zerah Woodford; East, part on land of said Payne, part on land of said Woodford, and part on land of Gideon Walker; South, part on passway, or avenue, and part on my own land; West, part on my own land, as the fence now is, and part on the eastern border of the pond, at high water therein, and part on the east bank of the [Roaring] brook...with the express understanding that the grantees...shall, at all times hereafter return into said pond, or suffer to accumulate therein, not less than one half the water which, at any time, may constitute said brook...and also to return the residue into the brook below the dam, at some convenient place, north of the south line of the above described and granted premises...[39]

The 20 acres transferred in the deed all lay to the east of Roaring Brook, and east of the former Seymour & Williams upper mill pond, but north of, and excluding, both of the former clock factory sites. Reference 1 (p. 294) mentioned a waterwheel on the property, but the deed does not say that. The parcel, bounded south partly by lands of Timothy Cowles, and partly by a passway (later known as Keyes Street), included the two old yellow factory boardinghouses that had been part of Solomon Langdon's former woolen mill complex.[40] With the firm's acquisition of the new property came the right to use half of the flow in the brook, subject to the requirement that the water be returned to the brook upstream of the next downstream user, occupying the former Seymour Williams & Porter and Seymour, Hall & Co. eastern clock factory building or case shop. It appears that the remaining half of the flow in Roaring Brook was being reserved for the use of an occupant of the former western clock factory building that had been destroyed in the fire, then or soon to be rebuilt.[41] In addition to this arrangement, on January 14 Cornelius R. Williams and the Prestons leased the right to build an additional dam north of their new mill site, partly in Unionville and partly in Avon, from William Griswold.[42]

In a now familiar pattern, Williams, Orton, Prestons & Co. mortgaged the entire property back to Timothy Cowles, for the same price, conditioned on payment of a note dated January 13, 1840:

...for three thousand three hundred dollars, payable to him [Cowles] with interest from date, in clocks...at the following times and manner, viz. three hundred on the first day of October 1840; two hundred and twenty five on the first day of April 1841, and the remainder, with the interest, on the first day of October 1841 [sic], then this deed shall be void.[43]

Thus Timothy Cowles remained deeply invested in clock making at Unionville.

During August 1840, Williams, Orton, Prestons & Co. brought suit against Daniel Tuttle (Case CC, Table 6A), formerly of Farmington, but at that time living in Ohio, on the grounds Tuttle had incompletely paid several debts he owed the firm. Briefly, between February 11 and June 12, 1839, Tuttle had purchased items, including ninety-six 30-hr. clocks at $4 each, six 30-hr. "Ogee front" clocks at $6, and ten wood 8-day clocks at $7, plus quantities of clock glass of various sizes, steel bells, pendulum rods, and clock keys,[44] all of which suggests he was organizing a peddling trip. Note the latest style 30-hr. ogee clocks (probably with stamped brass movements) priced at $6, while others, presumably the older style, 30-hr. wood movement half-pillar and splat types, were only $4, except the 8-day wooden clocks, which were $7. In addition, loose "looking glass plates" [mirrors] were a little more expensive than tablets, presumably hand-painted. Tuttle defaulted on his appearance in court; as a consequence, the court issued execution of its judgment in favor of the plaintiffs on March 25, 1841. It is interesting that the suit named Williams, Orton, Prestons & Co. partner Heman H. Orton as Tuttle's "agent or debtor."

Evidently, all went smoothly for a while, and on October 1, 1840, Williams, Orton & Prestons & Co. delivered the first installment of clocks owed to Timothy Cowles. In what appears to have been an unusual (but perhaps self-serving) personal endorsement, Cowles placed an ad two days later in the *Hartford Times*, offering clocks "of approved manufacture" for sale in Farmington (Figure 54).[45]

On February 8, 1841, Williams, Orton, Prestons & Co. initiated another suit (Court Case DD, Table 6A), this time against clock casemaker John Hunt of Plainville village (then part of Farmington), for failing to pay the firm for "615 Looking Glass Plates" [mirrors], valued at 16 cents each. The constable's attachment of Hunt's property included, in part:

...3 Clocks, 5 Clock Cases, 300 clock Cases partly finished, 300 clock Cases sawed out but in the rough, 500 clock Cases common or short ones, ¾ of Box Clock Weights, 6 Clock Faces, 46 Glass 10 by 10, Quantity of Glass 6 by 8, 72 Tablets, 36 C Bells, lot of Nob Turns, 2 Gross Brass Hinges, 12 Gross Screw Butts, 22 OG clock Sash, 7 unfinished Sash, 2 OG frames, 100 pr Top Blocks, 1 lot of venear Banding, 30 Hand Screws, 1 Screw Press, 1 hand Tool for Sticking OG &c, 15 Circular Saws, 1 Saw Cap, and 1 large Veneering Saw [sic].[46]

But Williams, Orton, Prestons & Co. was not the only creditor with a pending suit against Hunt. Following the listing was the statement: "...all the above described property was previously attached by three writs viz. Wm Crampton & Co.- Benjamin F Hawley & Wm Crampton & Co. as the property of the within named John Hunt..."[46]

On November 30, 1841, a few months after initiating the suit against Hunt, Williams, Orton, Prestons & Co. also brought a suit against Daniel F. and George M. Olcott [sometimes "Alcott"], both of Simsbury, CT (Case FF, Table 6A). The Olcotts had sold clocks for several Connecticut clock-making firms during the 1830s and 1840s, including C. & L. C. Ives of Bristol.[47] The matter in dispute was somewhat complex. The Olcotts had given Williams, Orton, Prestons & Co. a note in the amount of $438. Witnesses were summoned to testify regarding (among other things) the validity of a payment of $100.00, but their recollections conflicted. Nonetheless, the Olcotts subsequently made another payment of $352.22 on the note, so eventually the dispute centered on a remaining sum of only $7.00. The court, perhaps annoyed, rendered judgment in favor of the defendants. The plaintiffs, receiving nothing for their trouble, had to pay their own plus the defendants' court costs.

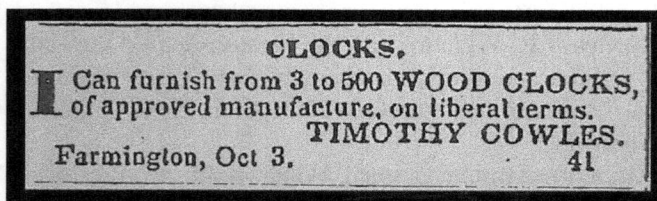

> **CLOCKS.**
> I Can furnish from 3 to 500 WOOD CLOCKS, of approved manufacture, on liberal terms.
> **TIMOTHY COWLES.**
> Farmington, Oct 3. · 41

Figure 54. Timothy Cowles's advertisement appearing in the Hartford *[Connecticut]* Times, *October 3, 1840.*

Table 7. Selected machinery and tool items from property of Williams, Orton, Prestons & Co.

Quantity	Description	Inventory Value
1	Iron Engine with Shaft, Belts & apparatus belonging	60.00
1	Wheel & 1 Pinion Engine	60.00
2	Lathes	40.00
1	Lathe & Motion	4.00
1	Iron Lathe & Motion	7.00
1	Iron Lathe, Motion & fixtures	5.00
1	Pinion Lathe	12.00
1	Drill Lathe, Bitts & Motion to Drive the frame	2.25
2	Iron Drill Lathes & fixtures with Shaft & Belts to Drive the same	10.00
1	frame with 3 Drill Lathes Motion & Bitts &c	10.00
1	Boxing saw frame & mandrell	4.00
1	Veneer Saw Frame & Mandrell	11.00
1	Splitting saw frame & Mandrell	6.00
1	Cutting off Saw	5.00
1	Saw frame, Mandrell Gages	6.00
1	Machine for Sawing Pinions	4.00
1	Saw plate Dies & Taps	5.00
2	Shafts & Bolts & planeing Machine	8.00
1	Planeing Machine & Motion	2.00
1	Anvill	6.00
1	Anvill in Smiths Shop	3.00
1	Smiths Bellows	5.00
1	Wire Straightener	1.00
1	Wire Cutter	10.00
1	Press	4.00
1	Press for Veneering	3.00
1	Stove & pipe	4.00
1	Kiln Dry	2.00
1	Burr Mandrell & motion	2.00
1	Jointer	1.00
1	Machine & wrench for Case Making	1.25
1	Pine Machine	1.50
1	Ogee Burr	.50

Regarding the ending date for the firm of Williams, Orton, Prestons & Co., according to the records of Court Case HH, mentioned above, the firm continued in business "to great advantage" "until about the 1st of January...[1842]," although the writ in Court Case GG, dated March 16, 1842, still referred to Williams, Orton, and the Prestons as "copartners." Nonetheless, in all four of the subsequent cases initiated by the firm in 1842 and 1843 (Table 6A, Cases II–KK), the court records named the partners and then described them as either "late copartners" or "late manufacturers and dealers." Therefore, in light of the evidence provided in

the records of the court cases in Table 6A, the firm of Williams, Orton, Prestons & Co. dissolved during the early part of 1842. What had happened?

It is odd that the court cases described above provided little indication that in fact the firm of Williams, Orton, Prestons & Co. stood on less than solid financial footing. On May 1, 1841, the firm, consisting of "Cornelius R. Williams, Heman H. Orton, Noah Preston and Eli D. Preston all of Farmington,...and Henry C. Bull, formerly of Farmington, now of Upper Alton, State of Illinois, partners..." assigned to Frederick W. Crum and Chauncey Rowe of Farmington, trustees:

Table 8. Clocks and clock parts belonging to Williams, Orton, Prestons & Co. as of May 1841.

Quantity	Description	Unit Price	Inventory Value
18	lbs. Sheet Brass	.33 1/3	6.00
264	lbs. Brass Chips	.07	18.48
96	lbs. Brass Chips	.04	3.84
45	lbs. Old Brass	.10	4.50
50	lbs. Scrap Brass	.14	7.00
80	lbs. Scrap Brass	.11	8.80
	Small tools for making brass Clocks		1.50
21	Brass 8 Day Clocks no wts or boxes	10.00	210.00
3	Brass 8 Day Clocks unfinished no wts or Boxes	9.00	27.00
9	8 Day Brass Clock Movements	3.00	27.00
18	pr 8 Day weights	.25	4.50
1	30 hr Brass Clock Movement		1.50
7	Ogee Clocks imperfect no wts or Boxes	4.00	28.00
125	Beveled Clocks unfinished no wts or Boxes	3.25	406.25
18	Beveled Alarm Clocks unfinished no wts or Boxes	3.62 1/2	65.25
47	Common Clocks no wts or Boxes	2.65	124.55
33	Common Clocks S Bells no wts or Boxes	2.75	90.75
7	Common Clocks imperfect no wts or Boxes	2.25	15.75
10	Clocks with Columns	3.75	37.50
711	estimated Clock Movements unfinished	.78	554.58
1300	Crown wheels		7.00
2000	Sett pinions 3.00 5000 pillars 2.50		5.50
	remant Lot of Wheel Stuff		4.00
1	Lot Clock Cord		8.00
28	Clock faces	.25	7.00
869	Clock faces	.20	173.80
4	faces	.12 1/2	.50
77	Clock faces unfinished	.10	7.70
92	Alarm Dials	.01 1/2	1.38
331	pr Small hands	.01 1/2	4.96
22	10 / 10 face glass	.05	1.10
40	8 / 8 face glass	.04	1.60
8	10 / 14 Look glass	.33 1/2	2.67
75	10 / 14 Look glass	.33	24.75
90	10 / 14 Painted Glass	.20	18.00
60	6 1/2 / 10 Painted Glass	.12 1/2	7.50
59	5 1/2 / 10 Painted Glass	.12 1/2	7.37
15	10 / 14 Painted Glass damaged	.04	.60

...all the personal property belonging to said Partnership...to them [the above], in trust, that [the trustees] shall take said property in their possession, and dispose thereof, and the avails thereof apply under the direction of the Court of Probate...in payment of all our Creditors, in proportion to their respective claims according as the same shall be established by Commissioners..."[48]

In addition, on that same date Cornelius R. Williams, Noah Preston, and Eli D. Preston, for simplicity sake referred to by the authors as "the three," each also made

a separate assignment of his "personal property and real Estate..." to trustees Crum and Rowe. On May 22 the *Courant* published a notice to "all persons interested in the estate of said company" to appear "(if they see cause)" on May 24 (Figure 55).

Meanwhile, on May 3, Edward Seymour and Virgil Goodwin (both former Unionville clockmakers, but probably signifying they were not employees of Williams, Orton, Prestons & Co.) were appointed appraisers of the personal and real estate belonging to both the firm of Williams, Orton, Prestons & Co. and that of "the three." However, on May 6 Crum and

302	pend balls	.03	9.06
14	Beveled frames		.75
75	Case backs	.03	2.25
	Refuse lot Sides & Backs	.25	
114	pieces Case Bottoms 150 tops	.01	2.64
332	turned Collumns	.01 ³/₄	5.01
36	Carved Collumns	.08	2.88
38	Carved Collumns	.06	2.28
26	Gilded Collumns	.09	2.34
200	Bronzed Collums	.12 ¹/₂	25.00
200	Painted Collumns	.08	16.00
1500	Estimated Collum Stuff		15.00
75	ft Collum Veneers	.04	3.00
1	Reffuse Lot veneers estimated		5.00
200	Wood frets	.60	
48	Bronzed frets	.04	1.92
82	Sett Bronzing	.12 ¹/₂	10.25
1	Lot pulley stuff		3.00
280	Com Bells	.02 ¹/₂	7.00
240	Steel Bells	.08	19.20
144	Bell Stands	.03	4.32
1000	Brass hinges	.00 ³/₄	7.50
400	pr Iron hinges for sash	.00 ¹/₂	2.00
1000	Sash turns		4.25
253	knobs & turns	.01 ¹/₂	3.79
265	knobs		1.98
170	pr wts	.15	25.50
489	pr Clock wts	.15	73.35
61	pr Common weights	.15	9.15
25	Alarm wts	.03	.75
22	Clock Boxes	.50	11.00
10	Clock Boxes	.37 ¹/₂	3.75
28	Weight Boxes	.22 ¹/₂	.70
82	parts wts Boxes	.02	1.64
1	Silver watch		23.00
1	Gold watch		16.00
1	English watch		6.00
1	French watch		3.00

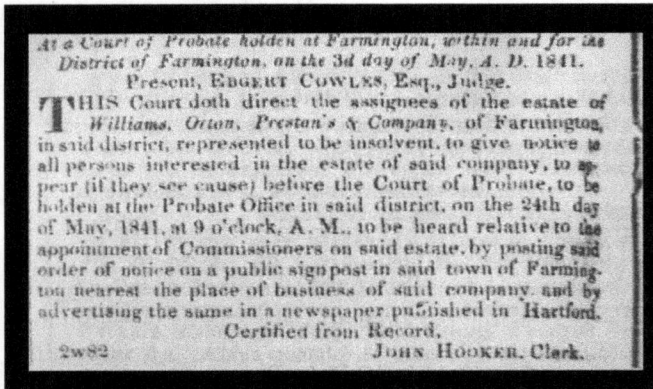

Figure 55. Notice to persons interested in Williams, Orton, Prestons & Co.'s assigned estate, Connecticut Courant, May 22, 1841.

Rowe resigned as trustees for all of the estates. In their place the Court substituted William Crampton and Ira Hadsell. On May 7, 1841, Cornelius R. Williams, Noah Preston, and Eli D. Preston quitclaimed the firm's real estate, along with the buildings thereon, for the amount

of $10 to Crampton and Hadsell. This was a normal action, in which the property of a bankrupt party was transferred to the trustees. Heman Orton was not named, because he was not an owner of the land. The property, previously described on January 13, 1840 (see above), was now described as follows:

> ...a certain piece or parcel of land situated in said Town [Farmington] and the buildings thereon consisting of four dwelling houses and one saw mill with the appurtenances thereto belonging, consisting of the right of taking water by a canal from the brook on land of William Griswold, and across land of George Payne, contains about twenty five acres; is bounded North on George Payne & Zerah Woodford, East on said Woodford & Gideon Walker, South on Timothy Cowles or avenue, & West part on said Cowles & part on [Roaring] brook.[49]

On May 13, 1841, inventories for Williams, Orton, Prestons & Co. and its partners' estates were presented

Table 9. Selected inventory items from property of Cornelius R. Williams, Noah Preston, and Eli D. Preston.

Quantity	Description	Inventory Value
1	Saw Mill, Shingle Mill, privilege and apparatus belonging	1450.00
	E.D. Preston's Dwelling	1000.00
	Noah Preston's Dwelling	950.00
2	Yellow houses	1550.00
24	Acres Land @20	480.00
44	Com Clocks without wts or Boxes @2.65	116.60
180	Com Clock Cases @.85	153.00
20	Com Cases – with Movements & faces @1.91	38.20
	Estimated parts in various stages for 750 Clock Movements @.37 ½	281.25
	Estimated parts in various stages of 1000 Clock Movements @.69	690.00
2200	Clock plates in kiln dry	20.00
1000	Sett pinion Stuff in kiln dry	12.00

to the Farmington District Probate Court. On the same day the Court empowered the trustees of the estates to sell some or all of the personal property for the firm of Williams, Orton, Prestons & Co., which consisted of "a great variety of Manufacturing machinery, mechanics' tools, Merchants' dry goods, Groceries, Horses, Household furniture, manufacturing stock. Products of manufacture &c. &c...." The actual inventory covered five pages with two columns each, or some 190 entries, with a total value of property amounting to $4,035.07,[50] providing much information of interest. A selected listing of the machinery and tools belonging to the firm of Williams, Orton, Prestons & Co. is provided in Table 7. Clocks and clock parts in the inventory of Williams, Orton, Prestons & Co. as of May 1841 are presented in Table 8. Finally, a summary of personal property belonging to the estates of Cornelius R. Williams, Noah Preston, and Eli D. Preston as of May 1841, extracted from the inventories, is presented in Table 9.

The records refer to the firm's "saw and shingle mill." Nonetheless, the inventory data provided in Table 7 depict a rather large, well-equipped, wooden movement clock factory. The appraisers recorded their inventory as they proceeded from the "Water Wheel, Bolts, Drums & Shafts, under Basement Story," then to the "Basement Story," "1st Story Turning Room," "In Case room," "1st fl[oor] outer room," "2d Floor," "2nd Floor North Room," finally ending up "in Garret," Although it is unclear from the records of Williams, Orton, Prestons & Co.'s assignment, the authors believe that the "saw and shingle mill" was used at least for clock casemaking, since (as we will see), the firm apparently continued to rent space in the old patent screw manufactory building of Norton, Cowles & Bidwell.

According to the Elisha Manross accounts in the collection of the Connecticut Historical Society, about 1841, Williams, Orton, Prestons & Co. sold a "leveller engine" and a lathe to the newly established clock-making firm of Manross, Wilcox & Co., as the latter firm was tooling up to produce brass movements. In addition to the above, from an examination of the information provided in Table 8, it was possible to gather evidence of brass movement clock making. Listed were some 550 lbs. of brass on hand, which seems to be a very

Figure 56A. Exterior of 8-day brass movement shelf clock by Williams, Orton, Prestons & Co., Farmington.

Figure 56B. Label (in poor condition) of clock shown in Figure 56A.

Figure 56C. 8-day brass movement of clock in Figures 56A and 56B. (Figures 56A–56C courtesy of The Unionville Museum.)

large quantity if the firm was not manufacturing brass clocks. Also listed was the item "Small tools for making brass Clocks." The inventory also reported one 30-hr. brass movement in stock,[51] although the authors do not believe it was made by the firm.

Documentary proof that Williams, Orton, Prestons & Co. did in fact produce brass movements was found in an official report dated July 15, 1839, addressed to the Comptroller of the State of Connecticut (then collecting data on manufacturers), in which Williams, Orton, Prestons & Co. stated that it expected to produce "about" 3,630 "wooden-wheeled clocks" and also 300 "brass-wheeled clocks" during the year ending September 1, 1839.[52] An example of a clock with an 8-day round front brass movement, manufactured "for George Marsh" was shown in Figures 19A and 19B in Chapter 1.[53] Another example, with the same or similar 8-day round front brass movement, this time with the label of Williams, Orton, Prestons & Co., Farmington, is presented here, in Figures 56A–56C. The clock's case is a modified triple-decker style. Its dial has a large opening in the center to allow viewers a glimpse of the clock's relatively costly brass movement.

In addition to the inventory of Williams, Orton, Prestons & Co.'s property, Table 9 presents a selected inventory of the personal properties of Cornelius Williams, Noah Preston, and Eli D. Preston ("the three"). The saw and/or shingle mill, and also the two yellow houses were listed here. For the three, the inventory consisted of "real Estate amounting to $5430.00 and personal Estate (clocks and parts of clocks, chiefly) amounting to $1420.21—Total $6850.21."

On May 14, 1841, during the early stages of his personal bankruptcy crisis and that of his firm, Noah Preston's wife Lucy (Marsh) Preston died, a little over

one month after having given birth to a baby daughter.[54] A bill dated May 16 from Moses Bart[h]olomew debited Noah: "to diging grave 1.00" and "to tending funeral with hors[e] to service 2.00."[55]

The probate processes on the insolvencies rolled on. On May 17, 1841, only three days after Lucy Preston's death, the Court appointed Sidney Wadsworth and Simeon Hart commissioners on both the firm's and the three partners' estates, "to examine and adjust the Claims of the Creditors of said Estate[s]…", and "that six months be allowed and limited…" for this process (Figure 57).[56] The commissioners presented their report of allowed claims to the Court on December 14, 1841, amounting to $4,821.81 for the company and $71.47 for the three. The allowed claims, mostly in the form of notes given by the firm, included outstanding sums due to the following: F. W. Crum ($40.32); Henry Mygatt, a merchant with a warehouse on the Farmington Canal at Plainville ($777.71); Levi Smith, then a wood movement maker of Bristol ($257.59); J. & A. Cowles ($110.94); Ira Bowen ($82.09); George Richards ($66.31); Ira Howe ($140.53); Cowles & Rowe ($436.28); Leonard Beecher ($468.22); Mark N. Porter ($161.57); Luther T. Parsons ($12.73); Walter S. Williams, a Hartford printer from whom Williams, Orton, Prestons & Co. purchased some of its labels ($15.00); Henry Bancroft ($78.54); Eber N. Gibbs ($206.47; Ellen Gibbs ($17.20); Hiram Apply ($24.62); Norman Warner ($130.52); Rufus A. Hitchcock ($202.68); Eliza Benton ($11.95); Charles Hitchcock ($43.52); Isaac P. Frisby [sic] ($88.42); Henry Thomson ($13.26); Charles L. Russel ($114.67); W. & C. Blakeslee ($173.05); and Phineas Curtis ($330.59). A number of these creditors were likely clock-making employees of the firm. In addition to the persons to whom Williams, Orton, Prestons & Co. was indebted by notes, the firm owed Solomon Whitman and "Gilbert & Williams" on book accounts. The appearance of "Gilbert & Williams" here, apparently William L. Gilbert and Cornelius R. Williams, each of whom had been a partner of George Marsh in the latter's firms of Farmington village, is interesting because it suggests Williams, Orton, Prestons & Co. had maintained a connection with Gilbert, who was about to commence large-scale production of 30-hr. and 8-day brass movement shelf clocks at Winsted. It is also interesting because an 8-day wooden movement shelf clock bearing the label of a firm by the name of "Gilbert & Williams", Montreal, L.[ower] C.[anada] has been reported.[57]

Returning once more to Frederick W. Crum's

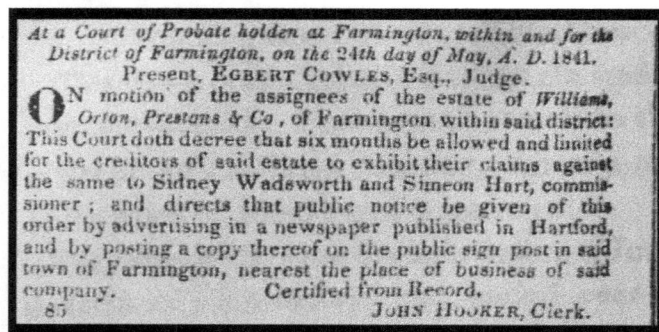

Figure 57. Notice to Williams, Orton, Prestons & Co. creditors of deadline to file claims, Connecticut Courant, June 12, 1841.

published recollections, the firm of Crum & Barber is said to have purchased a portion of Williams, Orton, Prestons & Co.'s business in 1841. Thereafter, Crum & Barber reportedly made clock movements, which Heman Orton cased. On the basis of (admittedly small numbers of) observed examples of Crum & Barber clocks, however, it is uncertain whether or for how long Crum & Barber actually made clock movements. Nonetheless, these statements are not only consistent with the 1841 assignment of Williams, Orton, Prestons & Co.'s estate but also suggest (in part) how the matter was resolved.

On May 22, 1841, a notice appeared in the *Connecticut Courant* (Figure 58). Dated May 10, it stated that Crampton and Hadsell, as Williams, Orton, Prestons & Co.'s assignees, had appointed Cornelius R. Williams as their agent responsible for working up the firm's stock of clocks in progress and for managing "the business of said company in furtherance of the settlement of said estate." Subsequently, on January 1, 1842, the *Courant* announced the sale at public auction of Williams, Orton, Prestons & Co.'s clock-making tools, machinery, general merchandise, and other property, scheduled for 9 A.M. on January 18 (Figure 59).

On the face of things, it seems reasonable to infer that the auction of January 18, 1842, marked the end of Williams, Orton, Prestons & Co.'s clock-making activities. However, neither the firm's land nor its buildings, nor its finished and unfinished clocks, were put up for sale at the auction. Therefore, while the business was being broken up and portions sold off, recalling that a combined total of about 2,000 or more clocks and movements in various stages of preparation were listed in the firm's and its partners' inventories, Williams, Orton, Prestons & Co. may have continued to send

out clocks under its own label for some time afterward. Indeed, evidence will be presented in Chapter 5 suggesting that both clocks and clock parts were sent out, presumably by the firm or its agents, for about two years following the auction sale of Williams, Orton, Prestons & Co.'s clock-making machinery.

In a deed dated March 2, 1842, Timothy Cowles quitclaimed, for $1.00, his interest in Williams, Orton, Prestons & Co.'s real estate to William Crampton and Ira Hadsell.[58] Shortly thereafter, on March 16, came the lawsuit of *Williams, Orton, Prestons & Co. vs. John O. Pettibone and Aurora Case of Simsbury* (Case GG, Table 6A), the latter presumably a clock sales firm. The partners in Williams, Orton, Prestons & Co. claimed Pettibone and Case owed them $500 in book debt. However, the suit was later withdrawn.

On May 6, 1842, William Crampton and Ira Hadsell quitclaimed the former Williams, Orton, Prestons & Co. property back to Williams and the Prestons, signaling the conclusion of the insolvency process.[59] On the same day, Williams and the Prestons executed another very complicated deed, which stated:

> ...for and in consideration of certain debts by us due and owing as hereinafter set forth do give, grant, bargain, Sell and confirm unto George Richards, Frederick W. Crumb, Henry Mygatt, Ira Howe, Ira Bowen, Timothy Cowles & Chauncey Rowe, partners under the name of Cowles & Rowe, James K. Camp, and Cowles, Deming & Camp, all of said Farmington; Leonard Beecher and Mark N. Porter both of Southington...and Phineas Curtis of Bristol... One certain piece of land at Unionville in Said town containing about twenty acres...[plus a

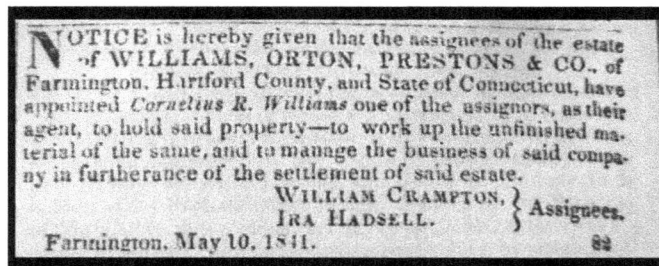

Figure 58. Notice of appointment of Cornelius R. Williams as agent for Williams, Orton, Prestons & Co.'s assigned estate, appearing in Connecticut Courant, *May 22, 1841.*

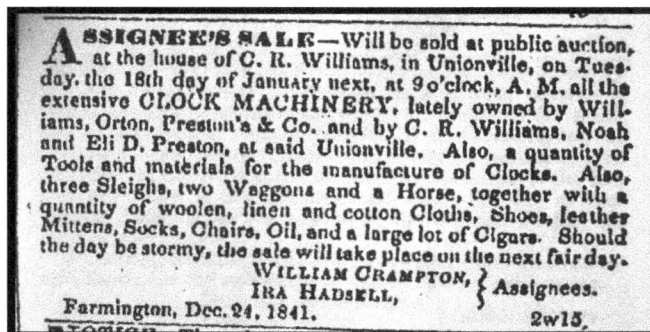

Figure 59. Notice of auction of Williams, Orton, Prestons & Co. machinery and tools, Connecticut Courant, *January 1, 1842.*

description of the property of Williams and the Prestons] and recorded in the 47th volume of Farmington Land Records, page 291 [i.e., Timothy Cowles's original deed to the partners of Williams, Orton, Prestons & Co., excluding Orton].[60]

So the factory property was deeded to a long list of creditors, "Provided that we [the sellers] shall well and truly, and within six months from date, pay to the said grantees [the creditors] severally the sums following with lawful interest thereon till paid viz to said [the creditors named above; the amounts to be paid to them (totaling $2,266.77) by notes dated concurrently, and signed by some combination of Williams, Orton, Noah Preston, or Eli D. Preston]. Then this deed to be void and of no effect, otherwise in force."[61] Although complex, this laid out a standard mortgage deed to the creditors, with Williams, Orton, Prestons & Co. scheduled to pay the debts by about November 6, 1842. This was the same deed examined in Chapter 1, showing that Orton, Preston & Co. had a connection with Cowles, Deming & Camp.

Returning to the records of Williams, Orton, Prestons & Co.'s assignment, the need for the above deeds will be made clearer. On May 6, 1842, the trustees reported to the Court that their final account was not ready, but they exhibited and settled their account for the estates of Cornelius R. Williams, Noah Preston, and Eli D. Preston. Figure 60 shows a photocopy of this account, taken from the middle of Volume 10, page 284, of the Farmington Probate Records. In Figure 60, in the Dr

(debit) column at left, the first item indicates, "To debts paid $73.31." The Commissioners' report for "the three" above indicates approved claims of $71.47. So these claims had been paid plus an additional $1.84. The second item is "To T[b] Cowles Mortgage raised $2,351.07." That allowed Cowles to quitclaim back the firm's real estate, described above. The next item on the list was "Cash to Agents (F. Crum: 20. CRW[ms] 60.00) $80.00." From this information, and that provided in the aforementioned *Connecticut Courant* ad of May 22, 1841, Frederick W. Crum assisted Cornelius R. Williams in finishing and disposing of the clocks in Williams, Orton, Prestons & Co.'s inventory (see Figure 60, third from last item). All other items represent the normal work of assignees William Crampton and Ira Hadsell, giving a total of $4,108.51. On the credit ("Cr") side, right column, is the total inventory $6,850.21, exactly as reported above. Finally, the total debit is subtracted from the total credit, giving a net credit of $2,741.70.

On May 20, 1842, the assignees' account for Williams, Orton, Prestons & Co.'s estate was made available to the Court. This is presented here as Figure 61, taken from the lower part of p. 284, Volume 10, of the Farmington District Probate Records. The debit column is again on the left. The first item, "To debts paid $4,946.09," is the amount reported by the commissioners above, $4,821.81, plus an additional $124.28 unexplained. The next seven items recorded normal assignees' business, some of which has already been mentioned above. It is interesting that a cash payment of $45.56 was made to J. & A. Cowles "for rent of Factory"! This bit of evidence confirms that a portion of Williams, Orton,

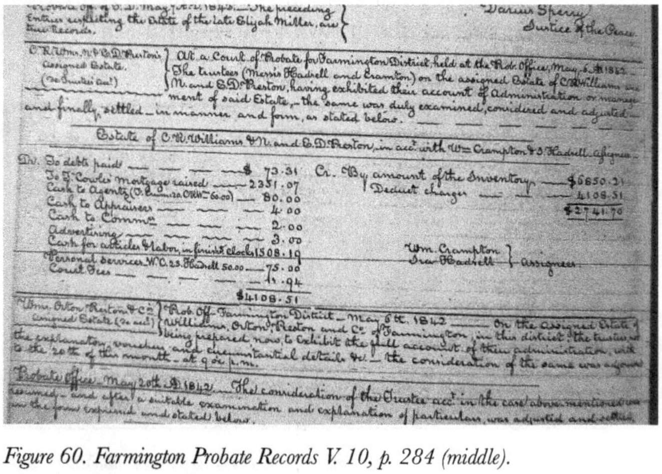

Figure 60. Farmington Probate Records V. 10, p. 284 (middle). (Courtesy Connecticut State Library.)

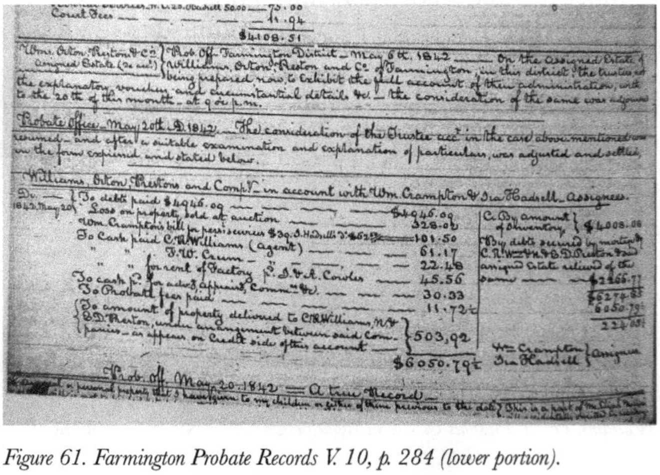

Figure 61. Farmington Probate Records V. 10, p. 284 (lower portion). (Courtesy Connecticut State Library.)

Prestons & Co.'s operations had continued within the (rented) former Patent Wood Screw Manufactory, of which J. & A. Cowles remained part owners.[62]

It is unclear what is meant by the last item in the trustees' combined report on the estate of Williams, Orton, Prestons & Co., which stated: "To amount of property delivered…." the amount of $503.92 not having been seen elsewhere. The total debits came to $6,050.79½. The first item on the credit side is "By amount of Inventory $4,008.08," which is $26.99 less than the $4,035.07 reported above and is not readily understood, although it is a small discrepancy. The next item is the amount of $2,266.77, the same sum that the partners issued notes to creditors for, all recorded in the long mortgage deed to the creditors, above. That allowed the credit side of the assignees' account to be increased by the same amount. The total debit amount was then deducted from the credits, giving a net credit balance of $224.05½.

On June 6, 1842, the Court ordered that the net credit of $224.05 for Williams, Orton, Prestons & Co. and the net credit of $2,741.70 for Williams and the Prestons both be turned over to the company. As we saw above, the firm's land and buildings were also returned to the company, but mortgaged for $2,266.77. This signaled the end of the bankruptcy proceedings. The assignees (trustees) had done well by the company.

On January 7, 1843, some of the firm's creditors, including Frederick W. Crum and the firm of Cowles, Deming & Camp, released Williams and the Prestons from the mortgage of May 6, 1842.[63] It is likely that Rufus A. Hitchcock and Henry Bancroft, witnesses to Frederick W. Crum's signature on the deed, were employees of Williams, Orton, Prestons & Co. In fact, Henry Bancroft was the husband of Amelia [sometimes "Aurelia"], one of Noah and Eli Dewey Preston's sisters.[64]

Returning once again to Court Case HH, Table 6A, on June 22, 1842, partners Cornelius R. Williams, Heman Orton, and Eli D. Preston of Farmington, and Henry C. Bull of Alton, IL, brought suit before the Hartford County Superior Court sitting in Chancery, against their fellow partner in Williams, Orton, Prestons & Co., Noah Preston. The substance of the partners' complaint against Noah Preston was as follows:

> …and the Petitioners say that during the whole of said term (to wit, form the 15 day of Dec[r] 1835 until the first day of January last past [i.e., 1842], the Deft. was Bailiff [meaning, "An overseer…or agent of an estate."[65]] [of] the Plffs and during that time had the Care and administration of a large number (to wit) of thirty thousand Clocks belonging to s[d] Company of great value to wit of the value of fifty thousand Dollars, to sell and dispose of the same to the best advantage for said Company and to render his reasonable account therof to the Plffs when he should be thereto required, yet the Defendant though often requested has never rendered his reasonable account thereof to the Plffs but has always refused & still refuses so to do, and has appropriated a larger sum over & above his proportion of the profits of said business (to wit) the sum of one thousand dollars to his own use, the Petitioners therefore pray your Honor to inquire into the facts aforesaid, and to require the said Noah Preston to render his account to the Plffs and to pay over to them such sum as may be due from him to the Plaintiffs in such amount as shou[ld be] just & reasonable, or grant such other relief as to right and justice appertains—[66]

On the same day, a constable of Farmington went forth and attached all of Noah Preston's right and interest in a piece of property, the metes and bounds of which clearly identify it as the above-referenced property of the firm of Williams, Orton, Prestons & Co. Noah defaulted on his appearance, which, being the defendant, meant he lost the case. As it happened, however, his partners withdrew their suit during the court's September 1843 term. We will return to this important matter shortly.

Meanwhile, during the fall of 1842, Cornelius R. Williams, Heman H. Orton, Noah Preston, and Eli D. Preston, all of Farmington, and Henry C. Bull of Alton, IL, "late partners in Company under the name of Williams Orton Prestons & Co.," brought suit against Oliver Weldon of Bristol (Case II, Table 6A). Weldon was a clockmaker who made some of his own wooden movements but had also on occasion purchased movements from Williams, Orton, Prestons & Co.[67] Subsequent to the court's judgment in favor of the plaintiffs, Deputy Sheriff Asa Bartholomew demanded:

> Several Sums contained in this execution… which he [i.e., Oliver Weldon] then & there neglected to pay and afterwards…at Said Bristol, I levied this execution on one case of Wood Clocks & four one horse Sleighs…and

took the same in to my possession…and posted up the same on the Sign post in…Bristol… and…I also set up a notice that said property would be sold…at the end of twenty one days at public Vendue… I offered said property for sale after causing the drum to be beat… [68]

Receiving no bidders, with the consent of Bristol Justice of the Peace Tracy Peck, the auction was moved indoors, to Foster's Tavern, where, in Bartholomew's words:

I…Sold said case of Clocks to Nathaniel Cramer…the highest bidder for the sum of nine dollars & fifty cents…and I also sold one sleigh to said Cramer…for the sum of fourteen dollars & Sixty two cents…one…to Samuel Jones…for five dollars and twenty five cents and…one sleigh to Billy S Hart for…three dollars & twelve & a half cents…and…one…to Erastus Foster for…ten dollars & fifty cents…amounting in the whole to thirty four dollars & ninety nine cents [sic]…[69]

Despite these efforts, the proceeds from the auction did not quite cover Weldon's debt.

On January 7, 1843, the same day on which (as mentioned above) several creditors of Williams, Orton, Prestons & Co.'s assigned estate released their interest in the firm's real estate and buildings on Roaring Brook, Cornelius R. Williams and the Prestons mortgaged its familiar property, referring to Timothy Cowles's deed of January 13, 1840,[70] to the First Ecclesiastical Society of Farmington for $500, to secure a note.[71] Noah Preston's son George and one Almira Cummings witnessed the deed. We can only speculate that the partners paid some of their creditors with the funds raised from the new mortgage. However, the remaining creditors were concerned (with some reason, because the mortgage of May 6, 1842, was not mentioned in this new mortgage, which stated only the words "free of all incumbrances having priority to the present"). Several of the numerous earlier mortgagees signed or signed with a statement below the official recording signatures, stating that this new mortgage was not to lessen the value of the earlier mortgage.[72] Nonetheless, this new mortgage did not default. Beside the deed in the Farmington Land Records is a note that reads: "This mortgage satisfied see Vol 62 page 221," signed by the town clerk. The authors have examined this last referenced deed, and found that the then-current treasurer of the First Ecclesiastical Society of Farmington quitclaimed the firm's property to Eli D. Preston on February 23, 1875.[73]

With its creditors satisfied and debts secured, the former firm of Williams, Orton, Prestons & Co. was winding down. On July 1, 1843, Noah Preston quitclaimed his interest in the firm's land, factory buildings, dam, and water privilege, to his brother, Eli D. Preston, for $1,300.[74] Clockmaker George Barber, and Caroline, the wife of Cornelius R. Williams, witnessed the deed. On the same day, Noah Preston transferred the one-acre parcel with his dwelling house and other buildings thereon, together with the privilege of taking water from the aqueduct on Cornelius R. Williams' and Eli D. Preston's land at his own expense, to Barber for $1,400.[75] Thus, it appears Barber, or more likely, the firm of Crum & Barber, had in some sense succeeded Williams, Orton, Prestons & Co., probably by purchasing "a portion of the business," as per Crum's published recollections. The deeds allowed Noah Preston's former partners to withdraw their suit against him.

Meanwhile, Noah Preston prepared to leave the state, albeit for the moment still technically as an employee of his former firm. On November 25, 1843, when the "late manufacturers and dealers under the name and firm of Williams, Orton, Prestons & Co." brought suit against William Daniels of Harwinton to recover $250 Daniels owed the firm on its books (Court Case KK, Table 6A), both Noah Preston and Henry C. Bull were identified as residents of Alton, IL. The documents initiating the suit contained an additional surprise: the listing of the firm's partners included the name of Frederick W. Crum! This information demonstrates Crum's close relationship with the firm of Williams, Orton, Prestons & Co., during (or even after) the conclusion of the legal process in the firm's bankruptcy. In August 1844 the suit against Daniels was withdrawn, suggesting that in the interim the debt was paid.

On July 19, 1843, Cornelius R. Williams quitclaimed his interest in a parcel with buildings on it, in the vicinity of Noah Preston's former home lot, to Eli D. Preston (possibly the latter's home site), for $125.[76] Four years later, on September 17, 1847, Cornelius R. Williams "now of Plymouth, Litchfield County and State of Connecticut," quit-claimed his interest in the rest of the property of the former firm of Williams, Orton, Prestons & Co., to Eli D. Preston. The property was now described as:

…containing about eighteen acres…with the buildings and appurtenances thereon. Bounded north on lands of Frederick W. Crum, George W. Payne & Eli D. Preston, East on lands of sd Payne, sd Preston, Eber N. Gibbs and George Barber, south on said Barber and avenue, West on Timothy Cowles. For further particulars see deed from Timothy Cowles, to sd Williams & Preston & Noah Preston, recorded in Farmington Records Vol 47th Page 290.[77]

Thus, by the fall of 1847 the firm of Williams, Orton, Prestons & Co. was substantially "unwound."

Although it is uncertain whether such a firm actually existed, on May 16, 1843, a land transaction between abutting property owners Zerah Woodford and George Payne referred to a parcel bounded west by "Preston Williams & Co."[78] Similarly, when George Payne mortgaged a nearby 40-acre parcel to Fisher Gay of Farmington on December 2, 1844, the parcel was described as bounded south "on highway in part, & Williams, Preston & Co., West on Frederick W. Crum..."[79] Williams and both of the Prestons had appeared on the town of Farmington's tax lists for the years 1841 and 1842 as co-owners of the former company land and buildings. The last year for which the three were taxed for business activity seems to have been 1841. During 1843, 1844, 1845, and 1846, only Williams and E. D. Preston owned and paid taxes on the former company properties. As of 1847 Eli D. Preston was the sole owner of three dwelling houses (his own plus the two former company boardinghouses), his own plus the former company land, and the company's former sawmill.[80]

In 1848, the year in which the Farmington Canal ceased operations to allow for construction of a railroad, Norton, Cowles, & Bidwell's local tax assessment no longer described the firm's industrial building as "the clock factory" but instead as "the old screw factory." However, on January 30, 1860, Eli Dewey Preston quitclaimed to John Treadwll Norton, guardian to James Lewis Cowles (1843–?), the oldest son of James Cowles (1795–1858):

> ...a certain...parcel of land situated in said Farmington in the village of Unionville, together with a factory building and a water wheel, connected with the same and all my use to said water. Bounded North and West on Heirs of James Cowles, East on Dwight Langdon and South on Farmington River containing one half acre more or less.[81]

By this deed Eli Dewey Preston relinquished his interest in the old Patent Wood Screw Manufactory land, building, and waterpower, to a descendant of one of its former owners.

For the clocks of Williams, Orton, Prestons & Co., an extant bill for items sold to Charles C. Jacobs during the summer of 1838 provides an indication of some of the case styles and finishes the firm offered:

Charles C. Jacobs

[18]38 [dr] to Williams Orton Prestons & Co.

[Ju]ly 11th to 24 Ogee front Clocks	6 ½	156.00
11 Steel Bells in the lot	25¢	2.75
24 Gilded Clocks	5 ½	132.00
		290.75
[Au]t 14th 24 Gilded Clocks	5 ½	132.00
25 Ex[tra] Clock[s]	6 3/8	159.37
45[?] Steel Bells		18.37
		310.74
		[error: 309.74]
		601.49
[Interes]t on 290.75 from July 11th 1839 6 Yrs 2 2/3 Mo		53.39
[Interes]t on 310.74 from Aug 14th 1839 4 Yrs 1 ½ Mo		76.15
		129.54
		731.03[82]

Note that Jacobs was very late in paying for the clocks. Similarly, a bill to a firm by the name of Tibbets & Owen of Altenbury, Perry County, MO,[83] for items purchased on September 5, 1838, reads as follows:

Messrs Tibbets & Owen

1838 to Williams Orton Prestons & Co. Dr

Sept 5th to 6 Brass Clocks		90.00
10 Ex 30 hour Clocks	6	60.00
24 Common	4 ½	108.00
5 Steel Bells		1.55
		259.55[84]

Like Jacobs, Tibbets & Owen was slow to pay its bill.

Figure 62A. Exterior view, 30-hr. wood movement shelf clock by Williams, Orton, Prestons & Co., Farmington, with modified triple-decker case.

Figure 62B. Label of clock shown in Figure 62A.

Figure 62C. 30-hr. wood movement of clock shown in Figures 62A–62B. See text. (Figures 62A–62C courtesy of Chris Brown; Gene Gissin photos.)

Figure 63A. Exterior view, 30-hr. wood movement shelf clock by Williams, Orton, Prestons & Co., Farmington.

Figure 63B. Label (stained) of clock shown in Figure 63A.

Figure 63C. 30-hr. wood movement of clock shown in Figures 63A and 63B. (Figures 63A–63C courtesy of Chris Brown; Gene Gissin photos.)

On the basis of the information provided in the two bills presented above, by 1838 Williams, Orton, Prestons & Co. had been selling at least four styles of 30-hr. wood clocks and also "Brass Clocks" (probably 8-day); the latter were the most costly. Presumably, the brass clocks were similar to (except perhaps for cases) that shown above in Figures 56A–56C, and to that shown in Figures 19A and 19B described in Chapter 1.

Surviving examples of 30-hr. wood movement shelf clocks produced by Williams, Orton, Prestons & Co. are not uncommon, and a few 8-day clocks have also been observed. A 30-hr. example is shown as Figures 62A–62C. Its case (Figure 62A), a modified triple-decker style, with full, smooth turned columns and a cornice top, is not infrequently seen in clocks of the firm. The label (Figure 62B) gives the clock's place of origin as "Farmington." Note "Prestons" in the firm name appears in the plural.

Both 30-hr. and 8-day wood movements made by Williams, Orton, Prestons & Co. evolved from those of its predecessors, Orton, Preston & Co., Marsh, Gilbert & Co., and George Marsh & Co. One difference, visible on the front plate of the movement in the present example (Figure 62C), is that the middle portion of the "Figure 8" access hole (upper left-hand area) has been substantially enlarged, a feature adopted by Williams, Orton, Prestons & Co. and other firms relatively late in the history of wood movements, probably to allow easier assembly. A second difference indicated in Figures 62B and 62C is that Williams, Orton, Prestons & Co. frequently bushed its movements with brass. This was not unique: a few other Connecticut firms also used brass bushings in their wood clock movements. However, reinforced bushings can be traced all the way from the Marsh firms (which bushed their movements with "ivory" or bone) to the later "internal" (i.e., not obvious in photographs) brass bushings of Orton, Preston & Co., which are also seen in some of the movements of Williams, Orton, Prestons & Co. clocks,[85] to the large, "external" brass bushings, visible on the winding arbor pivot holes of the movement of the present example (Figure 62C). Again, several other firms are known to have used such winding arbor bushings, another relatively "late" wood movement feature. No doubt they helped convince many customers that the clocks were visibly more durable than those made by competitors. Indeed, in his letter of June 30, 1839, mentioned above, Williams, Orton, Prestons & Co. clock customer H. P. Horton specifically requested brass bushed movements.

Figure 64. Stenciled and reverse painted upper tablet from another Williams, Orton, Prestons & Co. clock. Compare to upper tablet of clock shown in Figure 63A. (Lindy Larson photo.)

In contrast, the Terry family clock-making firms of Plymouth did not bush their movements with materials other than wood, and in fact spoke scornfully of the practice.[86]

A second example of a 30-hr. shelf clock, label, and movement by Williams, Orton, Prestons & Co., this time with carved columns and crest, is shown in Figures 63A–63C. According to its label (Figure 63B), this clock also originated at Farmington, and the firm name includes "Prestons." As in the previous example, the clock's movement is Type 5.153.[87] The bronze stenciling on the upper portion of the clock's tablet is unusual; upper tablets with similar stenciling have been observed in other examples of clocks made by the firm. See, for example, Figure 64.

Figures 65A–65C illustrate another Williams, Orton, Prestons & Co. clock. Its 30-hr. time-and-strike wood movement (Figure 65C) is similar to the previous examples, but has been lengthened to add an inboard alarm. Note the "extra" external brass bushing on the alarm winding arbor within the lower right-hand portion of the movement front plate. As in the above examples, the firm name on this clock's label includes "Prestons" in the plural and Farmington as the firm's location. A similarly painted upper tablet from a different example of a Williams, Orton, Prestons & Co. clock is shown for comparison, in Figure 66.

In addition to its 30-hr. and alarm offerings, Williams, Orton, Prestons & Co. produced a few 8-day wooden movement shelf clocks; an example is shown in Figures

Figure 65A. Shelf clock by Williams, Orton, Prestons & Co., Farmington, with 30-hr. wood time, strike, and alarm movement, exterior view.

Figure 65B. Label of clock shown in Figure 65A.

Figure 65C. 30-hr. time, strike, and alarm movement from clock shown in Figures 65A and 65B. (Figures 65A–65C courtesy of Chris Brown; Gene Gissin photos.)

Figure 66. Reverse painted upper tablet from another Williams, Orton, Prestons & Co. shelf clock. Note similarity to upper tablet of clock shown in Figure 65A. (Lindy Larson photo.)

Figure 67A. Exterior view, 8-day wood movement shelf clock made by Williams, Orton, Prestons & Co.

Figure 67B. Label of 8-day clock shown in Figure 67A.

Figure 67C. 8-day wood movement of clock shown in Figures 67A and 67B. (Figures 67A–67C courtesy of Chris Brown; Gene Gissin photos.)

Figure 68A. Label indicating the firm name of Williams, Orton, Preston's & Co. [italics added], Farmington.

Figure 68B. 30-hr. wood movement by Williams, Orton, "Preston's" & Co., in clock shown in Figure 68A. (Figures 68A–68B Rolf Taylor photos.)

67A–67C. Note again the appearance of "Prestons" and "Farmington" on its label (Figure 67B). The clock's movement, Type 4.22[88] (Figure 67C), resembles both the firm's own 30-hr. wood movements and the 8-day wood movements of its predecessor firms.

Eight-day brass movement shelf clocks by the firm have been discussed above. Again, the example of the clock with the unusual round front brass movement shown in Figures 56A–56C (above), is likely of Williams, Orton, Prestons & Co.'s own manufacture.[89] Although it is uncertain whether "Prestons" appears in the plural on the clock's badly disintegrated label, the location "[Farmin]gto[n]" (Figure 56B) is discernible.

It is interesting that some of the firm's clocks identifying "Farmington" as their place of origin are labeled "Williams, Orton, *Preston's* & Co." [italics added], with "Preston's" in the singular possessive, and the final letter "s" superscripted. Figures 68A and 68B illustrate the label and movement from one such example. The apparent change in wording is peculiar, and its meaning ambiguous, to the point of suggesting that the clocks with these labels were produced after Noah Preston relinquished his interest in the firm's land and buildings (and in the firm itself), to his brother, Eli D. Preston, on (or by) July 1, 1843.

All of the clocks illustrated so far have indicated "Farmington" as the place of origin on their labels. However, Williams, Orton, Prestons & Co. clocks with "Unionville" labels are occasionally seen. One such example, in a beveled case, is shown in Figures 69A–69C. Presumably, the firm's "Unionville" labeled clocks date to a later period than the "Farmington" examples. Indeed, it is likely that the designation "Unionville" came into more frequent use following the formation of the Unionville Ecclesiastical Society on July 15, 1839, and the organization in 1841 of the village's First Congregational Church, in which events a number of the clockmakers were active participants.[90] Furthermore, it is interesting that all of the firm's Unionville labeled examples examined by the authors

Figure 69A. Exterior view of 30-hr. wood movement shelf clock made by Williams, Orton, "Preston's" & Co., Unionville, in beveled case.

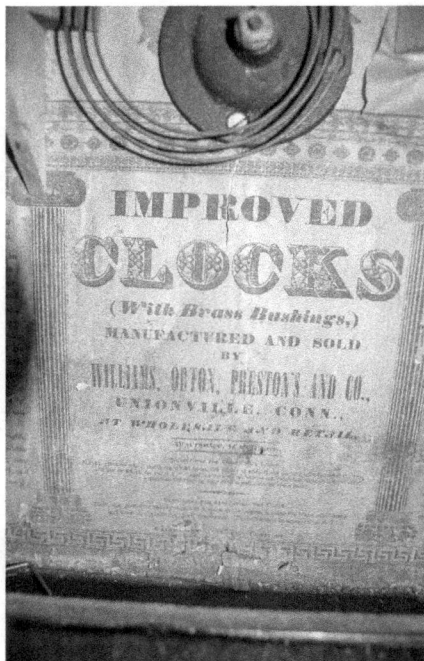

Figure 69B. Williams, Orton, "Preston's" & Co., Unionville label of clock shown in Figure 69A.

Figure 69C. 30-hr. wood movement by Williams, Orton, "Preston's" & Co., of clock shown in Figures 69A and 69B. Note count wheel retainer wire in 7 o'clock position and rectangular pendulum suspension stud. (Figures 69A–69C courtesy of private collection; MJD photos.)

clearly give "Preston's" in the firm name in the singular possessive (see Figure 69B). As we have seen, the label's printer, Walter S. Williams, at work in Hartford from 1840 to 1842, and then later, from 1844 to 1857, was a creditor of Williams, Orton, Prestons & Co.'s assigned estate. This information suggests that the clocks with Unionville labels were produced either during or after the assignment, drawing from the firm's inventory of movement and case parts. Another difference is the clock's wooden movement (Figure 69C), the second (and apparently far less common) of two basic 30-hr. versions considered to have been produced by Williams, Orton, Prestons [Preston's?] & Co.,[91] in which the count wheel retainer wire has been moved to the 7 o'clock position (Figure 69C). In light of the evidence, therefore, the Unionville labeled clocks would seem to date to about 1842, or perhaps even as late as 1843–44.

Returning once more to our consideration of the firm's history, thus far little has been said about Williams, Orton, Prestons & Co. partner Henry C. Bull (ca. 1812–85), the son of Benedict and Elizabeth (Carrington) Bull of Plymouth, who "learned the clock maker's trade."[92] Between 1835 and 1838, Henry C. Bull resided in Farmington, where he married Mary E. Cameron on November 4, 1835.[93] No deeds have been found for Bull in Farmington during the 1830s and '40s, but he appeared in the town tax assessor's records for the year 1838, for example, as the owner of "1 horse, 1 clock, and 1 watch."[94] Although in Illinois at the time, Bull was still described as "of Farmington" on August 29, 1840, when Williams, Orton, Prestons & Co. brought its suit against Daniel Tuttle (Court Case CC, Table 6A). However, as of the commencement of the firm's suit against John Hunt on February 8, 1841 (Case DD, Table 6A), Bull was residing in Alton, IL, where he and his wife made their home until at least 1860.

Henry C. Bull's removal to Alton is interesting because indications that Williams, Orton, Prestons & Co. maintained a "depot" there had surfaced previously, in the surviving letter from Edward N. Preston, grandson of Eli D. Preston, of February 25, 1927, mentioned above. Of interest at this point in our narrative is the portion of Edward N. Preston's letter that reads: "... They [Williams, Orton, Prestons & Co.] had a branch agency at Alton Ill[.] & the clocks were sent in parts but put together there..."[95] We will examine the firm's activities in Illinois in Chapter 5.

Remarkably, a letter has survived in which Henry C. Bull described in his own words the status of the firm's

business in Alton, IL, as of July 13, 1840, about ten months before Williams, Orton, Prestons & Co. assigned its estate in Connecticut for the benefit of its creditors. Addressed to "Messrs. Wms, Orton, Prestons & Co. [/] Union vill [/] Hartford County, Ct," the body of the letter reads as follows:

I wrote you some 4 weeks ago Respecting the Business & Prospects for selling Clocks for the year to come & also in Regard to hireing Thorp [undoubtedly, Eliakim Thorp, whom we will meet in Chapter 5] for another year & I have not Changed my mind on the Subject for it is of no use to hire men as long as the times Remain as they are. But Thorp thinks that he shall stay on the strength of the letter that you wrote last in which you gave it as the opinion of the Comp.[any] that they thought best for him to stay & work in the shop peddle & Collect & take Care of the Books But I have told him that we did not want him[.] But he thinks that the opinion of four is Better than one & that he shall stay for the Comp. now[.] I have no Objection to hiring Thorp if we wanted any one & should prefer him if he will work as Cheap as others but I am opposed to hiring him or any one at the present time.

I had a settlement with Thorp when his time was out & we owed him $275, for which I gave him our Note & in the settlement he Claimed $50 extra for keeping the Books which he says that Wms & N. Preston agreed to give him or that you would pay him what he thought was right, but he Concluded to take $45 & if Wms & N. Prestons say that if there was no such agreement that he would not Exact it he has not spent any Extra time in keeping Books & I think that he has had an easy time of it & that at the price that he had he could afforde to have done the wrighting without Extra pay but if you say that it is right then so be it. the times are harder than ever and the Collection[s] Come in slow. Mr. C.R. Woodford Came in last night and out of a lot of Notes of one thousand Dollars he had got but one Hundred, & every Dollar that we get the men want. I have sold the Waggon that we had of Warner for $170. fifty Dollars Cash in hand & the Balance I am to have the six of August. Hill & Afflick have not paid me for the horse that Hill bought of me when on East[?] do they send you any Monney

Have you seen P.M. Tillotson & has he paid you[?] he left har last winter for the East[.] Have Otis & Bishop of Indiana paid you yet[?] I have there Note for the Clocks that you sent them last fall it will be Due next <u>Oct.</u> The Amount is $400 with six per Cent interest

what do you think is best to do with the Clocks that we have on hand[?] I think that as the times are we had not better try to Peddle them at present But if the times take a turn for the Better by next fall I think that we might do well to Peddle them. I shall do the best I Can with them & wholesale them if I Can. But it is a hard Chance. Mr. Highley [sic, sometimes "Higby," one of the peddlers] will not go East this summer & if there is a Chance for selling he would like a job & he is first rate at the Business. you spoke of a Change that the Comp had. I think that it would be to Expencive one to send out here But if you Can get sutch waggons as the one that I got of Warner or larger I think that I could sell one or two at good prices. I wish that I had ten Thousand Dollars in mon-ney to send you...[96]

As an afterthought, written across the edge of the letter's last page were the words: "Mr. C.R. Woodford leaves for the East next Monday & he will bring all the newes [sic.]"

It is clear from Bull's letter that lingering impacts on the nation's economy from the Panic of 1837 were still being felt in the summer of 1840, even in such a remote location as Alton. However, it is unclear what may have constituted the "Change that the Comp had" Bull referred to in his letter, or what aspect thereof Bull thought would prove costly to implement in Illinois, or what news C.R. Woodford brought back to Connecticut. Nonetheless, having answered a few basic questions about the firm of Williams, Orton, Prestons & Co., we will take up these and other matters, and will encounter the peddlers Eliakim Thorp and Erastus Higby again in Chapter 5, in which we examine Williams, Orton, Prestons & Co.'s ill-fated Illinois clock making and sales venture.

Notes and References

1. Kenneth D. Roberts and Snowden Taylor. *Eli Terry and the Connecticut shelf clock*. 2nd ed. Fitzwilliam, NH: Ken Roberts Publishing Co., 1994.

2. Latter Day Saints, Family Search (online source) and report by Chris H. Bailey, dated September 6, 2006, research files of the American Clock & Watch Museum.

3. Town of Farmington, Tax Records, RG62, Connecticut State Library. In addition, according to extant accounts of the wood movement shelf clock-making firm of Hopkins & Alfred of Harwinton (ca. 1831–39) in the collection of the American Clock & Watch Museum, brothers Eli D. Preston and Gardner Preston purchased a few clocks from Hopkins & Alfred at various times between March and November 1835, but not afterward, suggesting that the men had been involved in a clock sales business shortly before Eli D. Preston moved to Farmington.

4. Mabel Hurlburt. *Farmington town clerks and their times.* Hartford, CT: Finlay Press, 1943.

5. *Williams, Orton, Prestons & Co. vs. Noah Preston.* Hartford County Superior Court Files, Box 96, RG3, Connecticut State Library.

6. *Samuel E. Judd vs. Williams, Orton, Prestons & Co.* Hartford County County Court Files, Box 430, RG3, Connecticut State Library.

7. *Bristol Herald,* July 3, 1890.

8. Town of Farmington, Tax Records, RG62, Connecticut State Library.

9. The differences in the assessments from year to year apparently reflect changes in the number of horses and wagons owned by the firm.

10. Town of Farmington Records, Box 12, RG62, Connecticut State Library.

11. Farmington Land Records (FLR), Volume (V.) 46, p. 527, October 16, 1837.

12. Research Activities & News. "Williams, Orton, Prestons & Co." *NAWCC Bulletin*, No. 264 (February 1990): 77–78.

13. Charles H. Preston. *Descendants of Roger Preston of Ipswich and Salem Village.* Salem, MA: Essex Institute, 1931.

14. FLR, V. 46, p. 528, April 23, 1838.

15. Harwinton [CT] Church Records, V. A., p. 15, Connecticut State Library.

16. Christopher Bickford. *Farmington in Connecticut.* 2nd ed. Farmington Historical Society, 2008.

17. William L. Warren manuscript collection, MS100987, Connecticut Historical Society.

18. H. P. Horton to Williams, Orton Prestons & Co., June 30, 1839; and Steven Colegrove to Williams, Orton, Prestons & Co., December 19, 1839, American Clock & Watch Museum.

19. Mary Jane Dapkus. "Hunt & Colegrove revisited." NAWCC *Bulletin*, No. 368 (June 2007): 321–322.

20. David D. Marsh. "Semi-Centennial Anniversary of the First Church of Christ, Unionville, 1841–1891. Unpublished manuscript, Connecticut State Library.

21. Clifford T. Alderman. *Images of America: Unionville.* Charleston, SC: Arcadia Publishing, 2010: 34.

22. Samuel Pepper. "History of Unionville." Unpublished manuscript, ca. 1924, Farmington Public Library.

23. Tax records for this firm were sometimes jointly listed with "Norton, Cowles, Bidwell & Co."

24. See, for example, FLR V. 45, p. 157, January 3, 1835, for evidence that Oliver Moore was a partner in the Patent Wood Screw Manufacturing Co.

25. Town of Farmington Tax Records 1833–1836, RG62, Connecticut State Library.

26. This may have been Rufus Stone, of the paper-making firm of Stone & Carrington. (See: Bickford, *Farmington*, 2008.)

27. See: Henry A. Castle. *The history of Plainville in Connecticut 1640–1918.* Plainville Historical Society, 1996: 35. See also Chapter 1 in the present work.

28. "CHARTER OAK...Prospectus." Julius Gay Papers, Box 3, Farmington Public Library.

29. Connecticut General Assembly Papers 1838–1839, Box 28, RG2, Connecticut State Library.

30. FLR, V. 47, p. 252, June 30, 1838–August 16, 1838.

31. FLR, V. 47, p. 290 (upper), December 28, 1839.

32. In addition to Bradley, another Unionville resident connected with clock making, Daniel B. Johnson, may have been the brother of Edward Seymour's wife Harriet (Johnson).

33. Roberts and Taylor. *Eli Terry,* 1994.

34. FLR, V. 47, p. 290 (middle), December 28, 1839.

35. "Index to Petitions for Naturalization Filed in New York City 1792–1989." Ancestry.com (online source).

36. "U.S. and Canada, Passenger and Immigration Lists, District and Port of New York." Connecticut State Library.

37. Town of Farmington, Abatements, Bills, Resolutions, etc., RG62, CSL.

38. See: FLR, V. 47, pp. 330–331, July 19, 1841; and FLR, V. 52, p. 154, July 13, 1850. E. K. Hamilton was still occupying the building as a hook and eye manufacturer at the time of the publication of Smith's *Map of Hartford County, Connecticut* in 1855.

39. FLR, V. 47, pp. 290–292, January 13, 1840.

40. Hurlburt. *Farmington town clerks*, 1943.

41. As of 1848 or earlier, this was David A. Keyes. See: FLR, V. 49, p. 424, August 2, 1848; and FLR V. 52, pp. 152–153, July 20, 1850.

42. FLR, V. 47, p. 304, January 14, 1840.

43. FLR, V. 48, p. 99, January 13, 1840.

44. *Williams, Orton, Prestons & Co. v. Daniel Tuttle.* Hartford County County Court Files, Box 441, RG3, Connecticut State Library.

45. *Hartford Times*, October 3, 1840.

46. *Williams, Orton, Prestons & Co. v. John Hunt.* Hartford County County Court Files, Box 442, RG3, Connecticut State Library.

47. See, for example, the Seth Wheeler Papers, RG69:47, Connecticut State Library.

48. Farmington Probate Records, V. 10; and Farmington District Probate Estate Papers, Connecticut State Library.

49. FLR, V. 45, p. 366, May 7, 1841.

50. Farmington Probate District Estate Papers, Connecticut State Library.

51. Ibid.

52. "Connecticut Comptroller's Return Schedules, Town of Farmington, Ct., 1839." Old Sturbridge Village, Sturbridge, MA collection on microfilm, Connecticut State Library.

53. Previously, on January 4, 1837, Williams, Orton, Prestons & Co. purchased a rather large quantity, 461 ½ lb, of rolled brass, from the insolvent estate of Jennes J. Woodward of Plainville village, apparently for use in the firm's own brass movement-making operations. However, the records of Woodward's insolvency, found in the Farmington District Probate and Estate Papers in the Connecticut State Library, suggest a possible source of some of the brass movements of the firm's predecessors. Woodward, who assigned his estate to trustees Harmanus M. Welch and Austin F. Williams before the Farmington District Probate Court on June 4, 1836, had been producing brass clock movements in a portion of Austin F. Williams's lumberyard in Plainville village. Woodward's creditors included the clock-making firms of Seymour, Hall & Co. and Orton, Preston & Co. Additional creditors included clock casemaker John Hunt of Plainville and even Timothy Cowles. Nothing is known about where Woodward went, or what he did, after the insolvency process on his estate concluded on March 30, 1838, with his creditors receiving 40 ½ cents on the dollar relative to the sums they were owed.

54. Charles H. Preston. *Descendants of Roger Preston of Ipswich and Salem Village.* Salem, MA: Essex Institute, 1931. According to this source, the couple's daughter, Lucy Adeline Preston, was born in Unionville on April 5, 1841.

55. Noah Preston Probate File, Illinois Regional Archives Depository, Carbondale, IL.

56. Farmington Probate District Estate Papers, Connecticut State Library.

57. "Research Activities & News." NAWCC *Bulletin*, No. 356 (June 2005): 359. The clock's 8-day movement, Type 4.111, was made by Jeromes & Darrow of Bristol. According to the information about George Marsh and the latter's firms presented in Chapter 1, prior to 1830, Marsh and William L. Gilbert had worked together at Bristol for Chauncey Jerome. Cornelius R. Williams became Marsh and Gilbert's partner about 1832. Note also that Jeromes & Darrow movements are very similar to those of the Marsh firms.

58. FLR, V. 45, p. 396, March 2, 1842.

59. FLR, V. 45, p. 413, May 6, 1842.

60. FLR,, V. 47, p. 352, May 6, 1842.

61. Ibid.

62. See: FLR, V. 47, p. 132, March 10, 1835; FLR V. 47, p. 32, November 2, 1833; and Town of Farmington Tax Records for the firm of Norton, Cowles & Bidwell, RG62, Connecticut State Library.

63. FLR, V. 45, p. 433, January 7, 1843.

64. Clifford Alderman personal communication to MJD, April 6, 2013; and Harwinton Vital Records, Connecticut State Library.

65. *Webster's Dictionary.* 2nd College ed. New York, NY: World Publishing Co., 1970.

66. *Williams, Orton, Prestons & Co. vs. Noah Preston.* Hartford County Superior Court Files, Box 96, RG3, Connecticut State Library.

67. Roberts and Taylor. *Eli Terry*, 1994.

68. *Williams, Orton, Prestons & Co. vs. Oliver Weldon.* Hartford County County Court, Executions, Box 533; and Hartford County County Court Files, Box 444, RG3, Connecticut State Library.

69. Ibid.

70. FLR, V. 47, p. 291, January 13, 1840.

71. FLR, V. 47, pp. 364-365, January 7, 1843.

72. Ibid.

73. FLR, V. 64, p. 221, February 23, 1875.

74. FLR, V. 45, p. 461, July 1, 1843.

75. FLR, V. 48, p. 433, July 1, 1843.

76. FLR, V. 45, p. 476, July 19, 1843.

77. FLR, V. 50, p. 123, September 17, 1847.

78. FLR, V. 48, p. 545, May 16, 1843.

79. FLR, V. 48, p. 546, December 2, 1844.

80. Town of Farmington Tax Lists, RG62:052, Connecticut State Library.

81. FLR, V. 54, p. 382, January 30, 1860.

82. Noah Preston Probate File.

83. William L. Warren Manuscripts, MS100987, Connecticut Historical Society.

84. Noah Preston Probate File.

85. Snowden Taylor. "Williams, Orton, Prestons & Co., Farmington, Ct." NAWCC *Bulletin*, No. 347 (December 2003): 788–789.

86. Roberts and Taylor. *Eli Terry*, 1994: 327–328.

87. Williams, Orton, Prestons & Co. was first identified as a movement maker, and its most common style of movements received the Type 5.153 designation, in the article by Snowden Taylor titled "Update on Terry-type 30-hour wood movements." NAWCC *Bulletin*, No. 249 (August 1987): 314–315. Subsequently, author ST's updated classification scheme (unpublished) proposes renumbering this movement as Type 7.153.

88. Bryan Rogers and Snowden Taylor. "Eight day wood movement shelf clocks--their cases, their movements, their makers." NAWCC *Bulletin* Supplement No. 19 (Spring 1993).

89. See: Snowden Taylor. "Research Activities & News." NAWCC *Bulletin*, No. 264 (February 1990): 76–78, for previous discussion and analysis that distinguished this movement from similar movements made by Silas B. Terry of Plymouth.

90. First Church of Christ, Unionville, Records, Connecticut State Library. It is uncertain based on label printer data alone whether the examples examined labeled "Unionville" date to a later period than those labeled "Farmington."

91. Snowden Taylor. Unpublished manuscript.

92. Francis Atwater (comp.). *History of the Town of Plymouth, Connecticut*. (Meriden, CT: Journal Publishing Co., 1895). According to Plymouth Cemetery Records, Henry Carrington Bull was born on October 29, 1812, and died August 24, 1885. He was buried in Plymouth's West Cemetery, alongside a number of siblings.

93. Town of Farmington Records, Box 12, RG62, Connecticut State Library; and Vital Records Index, Connecticut State Library.

94. Town of Farmington Records, Box 12, RG62, Connecticut State Library.

95. Snowden Taylor. NAWCC *Bulletin* No. 264 (February 1990): 76–78.

96. H[enry] C. Bull to Williams, Orton, Preston's & Co., July 13, 1840, private collection.

Chapter 5

William, Orton, Prestons & Co. in Illinois

Chapter 4 related the story of Timothy Cowles's acquisition of the clock factories on Roaring Brook formerly occupied by Edward Seymour's firms. This followed the execution in August 1838 of a court's judgment in Cowles's favor against the former partners in Seymour, Williams & Porter for unpaid debts. By 1843 Seymour, who had lost all of his investments in the former clock factory buildings, houses, and lands, was selling clocks in the frontier West for Cowles.[1]

On January 13, 1840, Timothy Cowles transferred a portion of the land once owned by Seymour, Williams & Porter, but excluding the former clock factory parcels and buildings, to Williams, Orton, Prestons & Co., with the latter firm mortgaging the property back to Cowles on the same day. With these transactions, Timothy Cowles began underwriting yet another clock-making firm in Unionville village.

During May 1841, Williams, Orton, Prestons & Co. assigned its estate for the benefit of its creditors. On January 18, 1842, the firm's clock-making tools and machinery were put up for sale at public auction. Shortly thereafter, Williams, Orton, Prestons & Co.

Figure 70. Portion of Map of Illinois and Missouri, by David H. Burr (London: John Arrowsmith, 1839), showing Alton, at the approximate center, where the Missouri River joins the Mississippi. Edwardsville, Madison County's seat, lies to the northeast. Macoupin County, in which Bunker Hill is located, lies to the north. The relatively small area encompassed by St. Louis, MO, as of 1839 is shown on the west bank of the Mississippi River, at center bottom.

dissolved, leaving many rather complex matters still to be settled. Among them was the firm's branch assembly and sales agency in Alton, Madison County, IL. We now examine the evidence relative to Williams, Orton, Prestons & Co.'s Illinois branch, and the tragic circumstances surrounding its closure.

Alton, Illinois

Although its nearby rival, St. Louis, MO, had long been a thriving port city on the west bank of the Mississippi River, Alton, located on the river's east bank in Madison County, IL, was only sparsely settled as of 1830. When the resolution of Black Hawk's War in 1832 brought an end to Indian hostilities, settlers (some of them well-educated New Englanders, including some from Farmington, CT), rapidly populated the area.[2] Located near the Mississippi's confluences with the Missouri and Illinois Rivers, Alton began to prosper as a steamboat depot serving remote interior areas. However, the murder of abolitionist publisher Elijah P. Lovejoy by a mob in Alton in 1836, and the subsequent outrageous acquittal of the crime's perpetrators by a local court, received national attention in newspapers. Consequently, Alton became perceived as an undesirable place in which to live and do business. These events, followed closely by the Panic of 1837, dealt a severe blow to the local community, and many of Alton's residents departed. It was only after more than a decade of decline that the town began to reestablish its reputation and economy.[3] A map showing Alton and its environs as of 1839 is provided in Figure 70.

With the nation's economy still suffering from the effects of the Panic of 1837, and Alton still mired in a depressed state of affairs, intense competition between Connecticut's clock manufacturers was not only forcing prices downward, but wooden clock movements were also rapidly becoming obsolete. About 1839 the Jeromes of Bristol introduced a 30-hr. movement stamped from locally rolled sheet brass. Although initially somewhat more expensive than wooden movement shelf clocks, new models containing these movements were selling well. It appears that part of Williams, Orton, Prestons & Co.'s strategy to remain competitive amidst such a challenging business climate had involved setting up an assembly plant at Alton.

Previously, a letter from Edward N. Preston of Unionville, grandson of Eli D. Preston, dated February 25, 1927, was mentioned.[4] Of particular interest here is the portion of Edward N. Preston's letter that pertained to Williams, Orton, Prestons & Co.'s activities in the American West. It reads:

They [Williams, Orton, Prestons & Co.] had a branch agency at Alton Ill[.] & the clocks were sent in parts but put together there[.] The State of Ill charged them of selling clocks without a license[.] The fight became so strong that my Grandfather had to send his local agent out there (This was Chester R. Woodford of West Avon then a part of F-n [Farmington] & he lived to be 107 years old & died 3 or 4 years ago), When Mr. Woodford got to Alton a real fight began[;] he had to procure a lawyer & he got Abraham Lincoln [who would become President of the United States some 20 years later] & he fought out the case [,] proved to the state that they needed no license for as the clocks were sent from Conn to Ill in parts boxes of wheels - faces hands minus the cases & put together in Ill[.] That constitutes the manufacture of the clocks, so won the case...[5]

Written two generations after the events occurred, the letter raised several questions: (1) Did Williams, Orton, Prestons & Co. actually have a branch agency in Alton, IL? (2) Was Chester R. Woodford really associated with Williams, Orton, Prestons & Co.? (3) Who was Chester R. Woodford? (4) Were clock movements really sent out to Illinois in parts and put together there? and (5) Did Abraham Lincoln actually handle such a case?

The answer to Question 1 was yes. An examination of the court cases summarized in Chapter 4 (Table 6A) revealed that Williams, Orton, Prestons & Co. partner Henry C. Bull removed from Farmington to Alton, IL, in 1839 or 1840. A surviving letter from Bull to his Unionville partners in Williams, Orton, Prestons & Co., dated July 13, 1840 (quoted in Chapter 4), not only documented the existence of the branch agency beyond any doubt but also made mention of a sales agent, C. R. Woodford.

Publication of Edward N. Preston's letter in the NAWCC *Bulletin* in 1990[6] produced a reply from James H. Whitaker, who reported an 8-day brass movement shelf clock in an ogee-style case with the label of "Williams, Orton, Preston's & Co. [sic] / Upper Alton, Ill."[7] Its brass movement was of the same type as in the clocks shown in Figures 19A and 19B of Chapter 1, and Figures 56A–56C of Chapter 4.[8] The existence of this clock, shown in Figures 71A–71C, seemed to suggest that Williams, Orton, Prestons & Co.'s depot at Upper Alton, IL, was an assembly plant. So, although the answer to Question 4 also seemed to be yes, Questions 2 and 5 remained largely unanswered.

Meanwhile, the authors became aware that according to a history of the town of Avon, CT, written during the late twentieth century, Chester Randolph Woodford and his uncle, Joseph Bishop, purchased clock parts in Connecticut and carried them to Illinois, but they found that the law there forbade the sale of imported products. When they assembled and sold some clocks, they were subsequently arrested. They retained a lawyer, Abraham Lincoln, to defend themselves. Maintaining that the clocks were not imported but "manufactured" in the state, Lincoln reportedly won the case.[9] Notwithstanding its similarity to Edward N. Preston's version of the story, there were still no primary sources connecting Bishop and Woodford with Williams, Orton, Prestons & Co., much less with Abraham Lincoln, in Illinois.

After some searching, author MJD discovered an important but undated document bearing on these matters. The circumstances that led to its discovery will be explained in the next section. For now, a transcript of the document in its entirety (preserving original spelling and grammar), is presented below:

Article of agreement made and entered into between Williams, Orton Prestons Co. of Farmington State of Connecticut & Bishop & Woodford of Avon Ct have this day united and formed a Copartnership for the purpose of selling Clocks & transacting such other business as may be mutually agreed upon by & between the parties under the name & firm of Bishop Woodford & Co- Now this indenture witnesseth that it is hereby declared & agreed by & between s^d parties that all contracts for Clocks goods Pedlars Wagons Harness & other expenses for commencing business & prosecuting the concerns of s^d Comp though made by either party for the account & benefit of s^d Co shall be for the account of the s^d Williams, Orton Prestons & Co & Bishop, Woodford & Co in the proportion as follows one half thereof for the s^d W.O.P

Figure 71A. Eight-day brass movement ogee-style shelf clock by Williams, Orton, Preston's & Co. [sic.], Upper Alton, IL, exterior view, door open.

Figure 71B. Label of clock shown in Figure 71A.

Figure 71C. 8-day brass movement by Williams, Orton, Prestons & Co., in clock shown in Figures 71A and 71B. (Figures 71A–71C courtesy of James Whitaker.)

& Co & one half for the s^d Bishop & Woodford & in like proportion all profit & loss thereon accruing or happening shall be had & received paid & sustain^d by the same W.O. P & Co & B & W and it is mutually agreed that the s^d Bishop & Woodford shall at the commencement of the concern put into the Co for to be disposed of in buying property for s^d Co or to be retained for expences & charges as they may think proper One Thousand Dollars and it is mutually agreed if s^d Co should need any more money for the commencement the s^d W.O.P & Co agree to furnish what shall be necessary for the prossecution of s^d Co. business until s^d B & W shall arrive & station themselves upon their pedling ground and it is mutually agreed by & between the parties that all monies necessary hereafter for the carrying on the business of s^d Co punctually furnished by both parties their equal proportion & in case either party shall advance more than his equal share or proportion above mentioned he shall receive from the other party the money back again & that for carrying on this contract no Commission shall be allowed to either of the parties nor any expense but what is necessary & the s^d W.O.P & Co & B & W mutually agree that each of the parties shall keep an account of all expenses of s^d concern relative to the business & deliver the same for examination to each & every one of the parties whenever requested so to do it is also agreed that after twenty one Months in rotation from 15th Sept next to the purchase of Clocks Goods Wagons Harnes hiring Pedlars & transacting all other business of s^d Co that there shall previous be a consultation of s^d parties if living & in health and if the business has proved profitable & s^d parties are agreed s^d Co is to continue the business another year or more, and it is also mutually agreed that if either of the above named parties (that is W.O.P & Co of the 1st part & B &W of the second part) shall at any time hold in his possession more than one half of the Notes papers & money belonging to s^d Co. directly or indirectly that on the application of any one party of the firm they shall deliver into his possession one half of all such property as they may have contracted of—it is also mutually agreed that the s^d B &W above named one or both shall at all times give their personal

attendance to the business of s^d Co among the Pedlars wherever they may be for the term of twenty one months from the time the Co starts out in the Western or Southern States[.] It is further understood & agreed that the s^d B&W are to furnish themselves with two good horses & two Wagons & harness to be driven by themselves while out attending to the concern of s^d Co. & it is also understood & agreed that the B&W are to receive 115 pr Month from the time they start from home until they return to be paid by s^d Co their necessary traveling expenses are also to be paid by s^d Co from the time they start until they return[.]

It is further understood & agreed that s^d W.O.P & Co are to furnish Brass & Wood Clocks at the following rates No. 1 of all kinds of Brass at 11.50 Extra Wood do 5.50 Common Wood 4.00 & it is also further agreed that s^d Clocks are to be paid for whenever the collection of s^d Clocks shall be made[.]

It is further understood & agreed that the s^d W.O.P & Co. shall furnish as many Brass & Wood Clocks as s^d B&W may want during 21 Months as aforesaid in the State wherever s^d BW&Co shall transact Business & it is mutually agreed that *s^d W.O.P&Co. shall Manufacture the s^d Clocks in the State where the s^d B.W.&Co shall transact business which State will be Illinois unless s^d Co agrees to remove to some other state* [italics added]--the s^d W.O.P &Co also agree that they will not sell Clocks that they manufacture in s^d State to any other firm or persons at wholesale that will interfer with BW&Co business in s^d State without B&W consent it is also understood that s^d B&W shall personally attend to the concerns of s^d Co & if s^d B&W should lose any time by sickness or any thing else their wages & expenses will be reduced it is also agreed that there will be added to the above named prices of Clocks the additional expense of freight & setting up the same - all money & other property is to earn interest from the time of delivery by either party {Signed } Bishop & Woodford {& Sealed} Williams Orton Prestons & Co.[10]

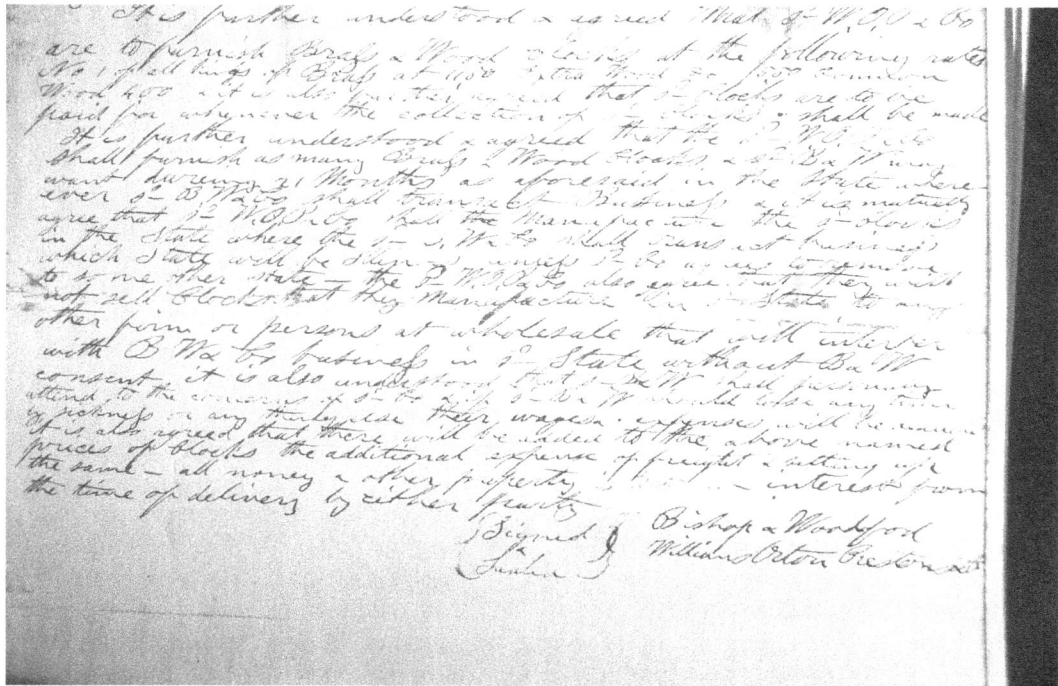

Figure 72. Portion of "Article of agreement" in which Williams, Orton, Prestons & Co. committed to manufacture "Brass & Wood" clocks for Bishop & Woodford in Illinois. (Courtesy of Illinois Regional Archives Depository, Carbondale, IL.)

A portion of the last page of the "Article of agreement" document, in which Williams, Orton, Prestons & Co. committed to manufacturing "Brass & Wood Clocks" in Illinois, is shown in Figure 72.

Under the terms of the agreement, Bishop and Woodford (first names not stated), became both Illinois agents and complex partners of Williams, Orton, Prestons & Co. Clocks were to be sent out in parts for "manufacture" in Illinois, answering Question 4 and also seeming to answer Question 2. Furthermore, based on the information provided in the agreement, and in a surviving "Schedule of Notes payable to Bishop, Woodford & Co.", with dates beginning on February 27, 1839, and extending as late as July 8, 1845, it seemed reasonable to assume the articles of agreement dated to about mid-September 1838, and also that a renewal beyond the original 21-month contract period occurred. Aside from this information, a surviving three-page list of notes payable taken in clock sales gave the counties, the names of the persons who gave the notes, the dates of the notes, and their due dates, on which the peddlers needed to return to collect on the notes. The notes originated in 37 counties, all apparently in Illinois. One hundred and seven notes were listed, totaling $2,446.61, although it is uncertain just how many dollars were actually collected.

From a family genealogy it was learned that Chester Randolph Woodford was born in 1814, to parents Chester Woodford (1782–1833) and Stella (Bishop) Woodford (1781–1887) of Avon [CT], and that he married Harriet Atwell Webster (1816–97), daughter of Ashbel and Esther (Bissell) Webster, in 1840. This source went on to state that Chester R. Woodford became a clock salesman, and later a farmer, and also that he had once been defended by Abraham Lincoln for peddling without a license in Illinois.[11] On December 11, 1921, the *Courant*, in its announcement of Chester R. Woodford's death at the age of 107, briefly repeated the story about Woodford's travels in the West, employing Abraham Lincoln as a lawyer to defend him in a civil action. These sources answered Question 3 above, relating to who Chester R. Woodford was, and confirmed that Joseph Bishop was his uncle. Question 5, asking whether Lincoln was actually involved, was still not satisfactorily answered, because the sources mentioning his involvement were secondary ones, written some 100 years after the events would have taken place.

The discovery of a foreclosure notice appearing in the *Alton Telegraph & Democratic Review* of August 25, 1848, provided a long hoped-for clue. The Alton Marine & Fire Insurance Co., the holder of a mortgage, was the plaintiff in a foreclosure suit that had been brought in 1845 on two properties located in Alton, IL. The borrowers in default were John Manning, Stephen Griggs, Joseph Bishop, Erastus Higby, Chester R. Woodford, Henry C. Bull, Horatio Bigelow, Nathan C. D. Taylor, and John T. Lusk. So Chester R. Woodford, Joseph Bishop, Erastus Higby, and others (some of whom [e.g., Stephen Griggs and John T. Lusk] had been early settlers and investors in Alton), were once co-owners of a property in which Williams, Orton, Prestons & Co. partner Henry C. Bull was also a co-owner, resolving any remaining doubt about whether Chester R. Woodford and Joseph Bishop were indeed the Bishop & Woodford in the agreement quoted above. Some eight pages of records of the foreclosure case contributed no additional information about the men's activities on the properties, and no information about possible clock-making activities at the sites.[12] The defendants defaulted on their appearance in court, and the properties were eventually sold to pay the debt. However, the court record did provide two bits of information consistent with the above-described primary and secondary source evidence of Williams, Orton, Prestons & Co.'s "branch" in Illinois: first, that Woodford and Bishop were "non-residents" of the State of Illinois; and second, that one of two parcels of land being foreclosed on was:

> ...situated in Upper Alton- commencing at the South East corner of William Harrison's land on Main Street- thence South on the line of said Street- Five Rods thence parallel with the south line of William Harrison's Lot- Ten Rods thence North parallel with the West line of Main Street- Five Rods and thence East along the South line of said Harrison's Lot to the place of beginning...[13]

The record contained no mention of any buildings on the property, which was evidently located on Main St. in Upper Alton. However, for reasons to be explained later, the authors believe the clock factory was rented and was not located at either of the sites that were the subject of the foreclosure case.

The Erastus Higby Case

Despite the discovery of the foreclosure case connecting Chester R. Woodford, Joseph Bishop, and Henry C. Bull with Williams, Orton, Prestons & Co. in Illinois, the answer to Question 5, asking whether there actually was a court case involving Chester R. Woodford and attorney Abraham Lincoln, remained unanswered. Subsequently, author MJD contacted the Illinois Regional Archives Depository at Springfield, IL, to inquire about additional court cases possibly involving Chester R. Woodford, Joseph Bishop, or any of the other the men named in the Alton Marine & Fire Insurance foreclosure case, perhaps with Abraham Lincoln as their attorney. She received the following reply:

> In response to your request, we searched the Lincoln Legal Papers and located the case *People ex rel Fishbourne v.* [Erastus] *Higby* (Marshall County Circuit Court), copies [44 pages] enclosed. The original case proceedings were in the Justice of the Peace dockets, which were not required by law to be filed of record. According to the case summary, Fishbourne informed the JP that Higby, a clock peddler, had failed to display a license. Fishbourne, in the name of the people of the state, sued Higby to collect the fine. The JP fined Higby $50, and Higby appealed to the circuit court. The court affirmed the JP judgment, and Higby retained Stephen Logan, Lincoln's senior partner and appealed to the Illinois Supreme Court. Apart from his association with Logan, there is no indication that Lincoln was directly involved in this case. We found no other relevant cases for the associated names you provided to us.[14]

Forty-four pages of documents accompanied the above message relative to the case titled *The people [of the State of Illinois] ex rel. Fishbourne vs. Higby*. From these documents it was learned that an Illinois law dating to the year 1835 required clock peddlers to obtain licenses. Furthermore, quoting from a summary of the case before a local Justice of the Peace:

> Be it remembered that on the trial of this cause, the Plaintiffs proved that the Defendant on or about the 20th Day of June AD 1839 Was engaged in the County of Marshall in peddling Clocks And for the said Defendant about that

time sold a wooden Clock in said County to One Julius H. Long for the Sum of Twenty dollars Without having or exhibiting a license to sell Clocks in Said County, and there upon the Said defendant proved that the Clocks which [he] carried about to Sell and also the clock sold by him aforesaid Were Manufactured in Alton in the State of Illinois and that Said defendant Sold as agent of the Manufacturers [thereby implying, but not proving, that this was a requirement of the law]. And there upon the Court gave judgment in favor of the plaintiff for the Sum of fifty dollars One half [to] the use of Abraham S. Fishburn the informer on this cause and the other half to the use of Said County of Marshall to which decision and

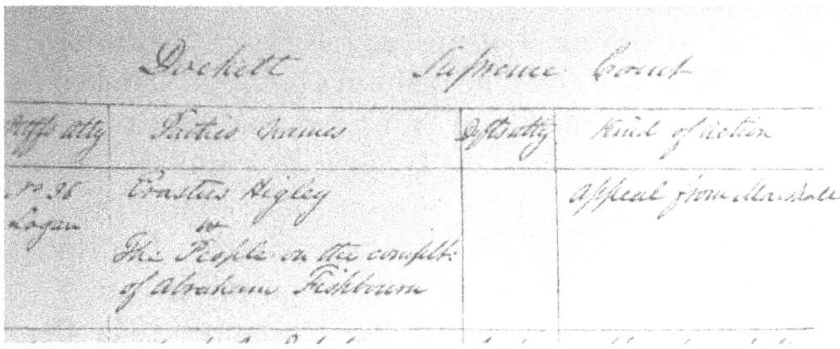
Dockett Supreme Court

Pltff atty	Parties names	Defts atty	Kind of action
No 35 Logan	Erastus Higley vs The People on the complt of Abraham Fishburn		Appeal from Marshall

Figure 73A. Example of Illinois Supreme Court "Dockett" for the Erastus Higby case, in which Higby's name seems to be spelled "Higley." Note name of plaintiff's attorney at left: "Logan".

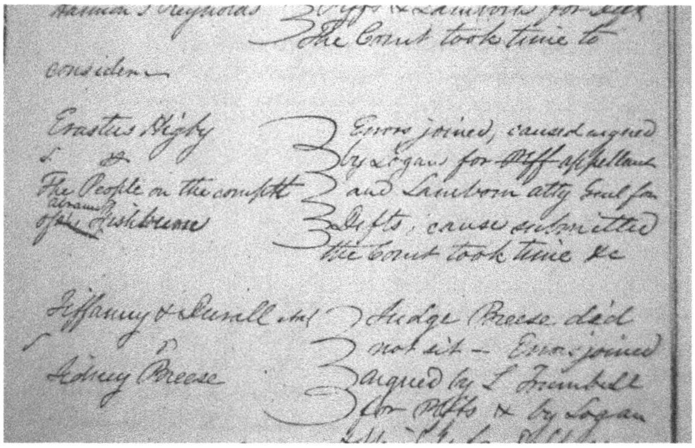
The Court took time to consider

Erastus Higby v. The People on the complt of Abram Fishburne — Error joined, cause argued by Logan for Pltff appellant and Lamborn atty Genl for Defts, cause submitted the Court took time &c

Tiffany & Sewell v. Sidney Breese — Judge Breese did not sit – Error joined argued by L. Trumbull for Pltffs & by Logan

Figure73B. A second Illinois Supreme Court docket example for the Erastus Higby case, in which the plaintiff's last name appears as "Higby." (Figures 73A–73B courtesy of Illinois Regional Archives Depository, Carbondale, IL.)

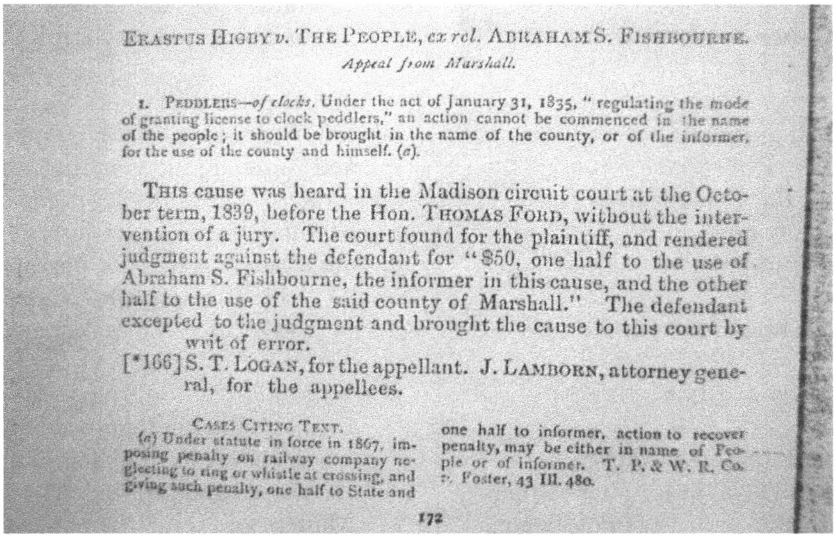
ERASTUS HIGBY *v.* THE PEOPLE, *ex rel.* ABRAHAM S. FISHBOURNE.

Appeal from Marshall.

1. PEDDLERS—*of clocks.* Under the act of January 31, 1835, "regulating the mode of granting license to clock peddlers," an action cannot be commenced in the name of the people; it should be brought in the name of the county, or of the informer, for the use of the county and himself. (*a*).

THIS cause was heard in the Madison circuit court at the October term, 1839, before the Hon. THOMAS FORD, without the intervention of a jury. The court found for the plaintiff, and rendered judgment against the defendant for "$50, one half to the use of Abraham S. Fishbourne, the informer in this cause, and the other half to the use of the said county of Marshall." The defendant excepted to the judgment and brought the cause to this court by writ of error.

[*166] S. T. LOGAN, for the appellant. J. LAMBORN, attorney general, for the appellees.

CASES CITING TEXT.

(*a*) Under statute in force in 1867, imposing penalty on railway company neglecting to ring or whistle at crossing, and giving such penalty, one half to State and one half to informer, action to recover penalty, may be either in name of People or of informer. T. P. & W. R. Co. *v.* Foster, 43 Ill. 480.

172

Figure 73C. A portion of the printed summary of the Erastus Higby case appearing in Vol. 4 of the Reports of cases argued and determined in the Supreme Court of the State of Illinois, *by J. Y. Scammon (Philadelphia, PA: J. Kay, Jr. & Brother, 1841). Note spelling of Higby's last name.*

judgment of the Court the defendant excepts and prays the Court to Sign And Seal this his bill of exceptions which is done.[15]

The above exception, or appeal, continued before the "Marshall [County] circuit court at the October term, 1839," where the previous judgment of $50 was reaffirmed, "And likewise the Costs and Charges... Whereupon the Defendant demands an appeal to the Supreme Court of the State of Illinois Which is granted herein," provided a bond of $150 was posted. The document, attested to be a true copy of the proceedings, bore the date November 29, 1839.

Another important document, headed with the case name *Erastus Higby vs. The People on complaint of Abraham Fishburne*, plus "In the Supreme Court of the State of Illinois at December Term 1839" and "On appeal from Marshall County" provided the following summary of Higby's argument before the Illinois Supreme Court:

> The Appellant comes and says that in this record, proceeding and Judgment of the Circuit Court for Marshall County Manifest error hath intervened to the prejudice of the appellant in the following particulars viz
>
> 1st The Court erred in rendering Judgment against the Appellant
>
> 2nd The Court erred in refusing to dismiss the case on motion of appellant
>
> 3rd The Court had no jurisdiction of the case
>
> 4th This Judgment of the Court ought to have been for the Defendant appellant Wherefore the appellant prays that said Judgt be reversed
>
> [Signed] Logan for appellant
>
> Dated Dec[r] 6. 1839[16]

On the basis of information provided elsewhere in the case records, "Logan" was indeed none other than Stephen T. Logan (1800–80), an Illinois lawyer, judge, and legislator. It should be noted that Logan and Lincoln had not yet formed their law partnership as of December 1839, when the Illinois Supreme Court first heard Higby's appeal. However, the case remained

unresolved when Logan and Lincoln became copartners during the spring of 1841.[17]

The Illinois Supreme Court docket for the June term 1840 recorded the Higby case as "Continued." Likewise, the case was continued through the December 1840, February 25, 1841, and July 1841 terms. One of the handwritten case dockets is illustrated in Figure 73A, in which Higby's last name could be interpreted as "Higley." Note "Logan" is identified as the plaintiff's attorney. A second handwritten docket, shown in Figure 73B, in which Higby's last name seems to be spelled "Higby," provides some information about the proceedings. A portion of the printed summary of the case that appeared in Volume 4 of the "*Illinois Reports*" (shown in Figure 73C), also gives the plaintiff's name as "Higby." On July 9, 1841, the Court "Ordered that this cause be placed at the foot of the docket of cases pending in this Court." The July 23, 1841, session "ordered that this cause be continued..." and the docket for the "December Term 1841" was also marked "Continued." On an unspecified date, but clearly in late December 1841, the court commented on the Higby case as follows:

> ...on this day came again the said parties and the errors being joined this cause was argued to the Court on the part of the appellant by S.T. Logan and on the part of the appellees by Lamborn Atty General and the cause was submitted- and the Court not being sufficiently advised took time to consider-.[18]

Another presumably very late docket in the case, labeled "Supreme Court December Term 1841" was marked "Agreed judgment reversed..." A docket for "Monday 17th Jany 1842 is marked "Reversed - opinion to be written by the Chief Justice." Finally, on the same day, on a different sheet of paper, the court commented on the case *Erastus Higby v The People & on the complaint of Abram Fishborn; Appeal from Marshall* as follows:

> On this day came again the parties aforesaid by their attor[nies] aforsaid and the court having diligently examined and inspected as well the record and proceedings aforesaid, as the Matters and things therein assigned for error, and being now Sufficien[tly] advised of and concerning the premises, are of opinion that in the record and proceedings aforesaid, and in the rendition of the judgment aforesaid, there

is Manifest error, therefore it is considered by the Court, that, for that error and others in the record and proceedings aforesaid; the judgment aforesaid be reversed, annulled, set aside, and wholly for nothing esteemed. And it is further considered by the Court that the appellant do recover against the appellee his costs by him in this behalf expended and that he have execution therefor.

> Opinion to be written by Chief Justice [William] Wilson.[19]

For reasons unknown to the authors, the summary opinion in the case of *Erastus Higby v. The People, ex rel. Abraham Fishbourne*, which overturned the lower courts' rulings against Higby, was neither affirmed, summarized, nor published until December 19, 1843, during the December 1843 term of the Illinois Supreme Court. The 44-page packet received by the authors contained three essentially identical court opinions, proposed by Illinois Supreme Court Justice S. H. Treat, but only the portion explaining the court's reasoning needs to be examined here. As described above, a clock peddler (such as Erastus Higby) was required to pay a penalty of $50 if he failed to show his license when offering a clock for sale. An informer (like Abraham Fishbourne, in this case) could report the failure to a local Justice of the Peace. The latter could impose the $50 penalty, which was to be divided equally between the informer and the County (such as Marshall County, in this instance). Quoting Justice Treat, in part:

> This is a qui tam action, the penalty going to the informer and the county. The state has no interest in [gets no part of] the recovery. The statute not authorizing the suit to be instituted in the name of the people [of the State of Illinois], it was improperly brought, and the [circuit] erred in not dismissing it. The action should have been brought in the name of the informer, or in that of the county; most properly in the name of the informer, for the use of himself and the county...[therefore,] The judgment of the circuit court is reversed.[20]

So Erastus Higby (with the help of Abraham Lincoln's law partner, attorney Logan) won the Illinois Supreme Court appeal on a technicality, rather than strictly on the basis that the clocks were produced in Illinois. Although according to at least one Lincoln scholar, the future U.S. president had some involvement in *all* of the cases brought before the Illinois Supreme Court in which his firm had been retained (specifically including the Erastus Higby case);[21] again, there is no evidence in the records reviewed by the authors of Lincoln's direct involvement. However, absent any other known court case in which Lincoln defended Chester R. Woodford directly, the details of the Erastus Higby case (in which Higby seems to have been an employee of Bishop, Woodford & Co.), are otherwise consistent with the two-generations-later memory of Eli Dewey Preston's grandson, Edward N. Preston, and with the recollections of Chester R. Woodford, whose daughters were said to have accidentally burned several letters to Woodford written by Abraham Lincoln, "during a spring attic cleaning spree."[22]

During the course of reviewing documents related not only to the Erastus Higby court case but also other primary source documents related to Williams, Orton, Prestons & Co.'s Illinois branch (both handwritten and printed), the authors came to believe that the man whose first name was "Erastus" and whose last name seemed to have been variously spelled as "Higly," "Higley," "Highley," or "Higby" was one and the same person: the defendant in the court case *People ex rel Fishbourne v. Higby*, whom we have chosen to refer to herein as "Higby." For example, this seems to have been the same man Henry C. Bull was referring to when, on July 13, 1840, he remarked in his letter or report from Alton to the firm's headquarters at Unionville: "Mr. Highley will not go East this summer & if there is a Chance for selling [clocks] he would like a job & he is first rate at the Business..."[23] Nonetheless, additional information raises questions about this interpretation. According to a family genealogy entitled: *The Higleys and Their Ancestry*, by Mary C. Johnson (New York, NY: Grafton Press, 1892), an Erastus Higley born in Simsbury [CT], on September 8, 1812, the twelfth child of parents Seth Filer and Naomi (Holcombe) Higley, who died unmarried in Illinois on August 14, 1847, at the age of 35, may have been the clock peddler in question. According to this source, Erastus Higley had once been employed as a traveling salesman for a "Bristol, Rhode Island manufacturing firm" and owned considerable property in Illinois. One of Higley's brothers had once been a clock peddler in the South, and another brother reportedly lived for a time at Nauvoo, IL. However, the authors have been unable either to confirm the information in the family genealogy, or to connect the man described therein with Williams, Orton, Prestons & Co.'s Illinois branch. Thus, while it appears that after the conclusion of the case, Higby or Higley remained in Illinois, some uncertainty about the identity of this mysterious peddler remains.

Meanwhile, at Farmington, CT, on June 25, 1841, Justice of the Peace John Hooker issued a summons and writ of attachment in a lawsuit brought by Austin F. Williams "of Farmington" (Timothy Cowles's son-in-law), and co-plaintiffs John D. Camp and George W. Abbe, doing business together in New York City under the firm name of Williams, Camp & Abbe. The defendants in the suit, Joseph Bishop and Chester R. Woodford of Avon; Cornelius R. Williams, Heman H. Orton, Noah Preston, and Eli D. Preston of Farmington; and Henry C. Bull of Alton, IL, were described as "doing business at said Alton and elsewhere" under the firm name of Bishop, Woodford & Co.[24] According to extant records of the case, Bishop, Woodford & Co. had given Williams, Camp & Abbe a promissory note for $691.07 on February 16, 1841, for "value received." The note had been payable in four months from the date of issue at the Hartford Bank, but June 16 had come and gone with no evidence of payment. As a result, on June 25, a house, barn, and a shop situated on a one-acre parcel, plus an additional three-acre parcel of land in Avon, both belonging to Chester

R. Woodford, and the homes of Eli D. and Noah Preston in Unionville, were all placed under attachment. This was less than two months after Williams, Orton, Prestons & Co., Williams, and the Prestons, assigned their respective estates for the benefit of their creditors. No further information is available concerning the suit or its outcome.

The Death of Noah Preston

In Chapter 4 we learned that Noah Preston's wife, Lucy (Marsh) Preston, passed away on May 14, 1841, and that during the summer of 1843, Noah Preston departed from Connecticut for the American West. At that time, Noah's personal bankruptcy proceedings, and those of the firm of Williams, Orton, Prestons & Co., were nearing closure. Noah and his partners were also waiting for Illinois Supreme Court Chief Justice Wilson's opinion, reversing the Circuit Court's decision in the case of *Erastus Higby v. The People, ex rel. Abraham Fishbourne,* to be written. Although the firm of Williams, Orton, Prestons & Co. no longer officially existed after the spring of 1842, the available information about Noah's activities in the West suggests that the firm (or a successor) continued to ship both finished clocks and clock parts to Illinois over the course of the next two years.

On the basis of remarkably detailed information provided in Noah Preston's Illinois probate file (see discussion below), after leaving Connecticut, Noah

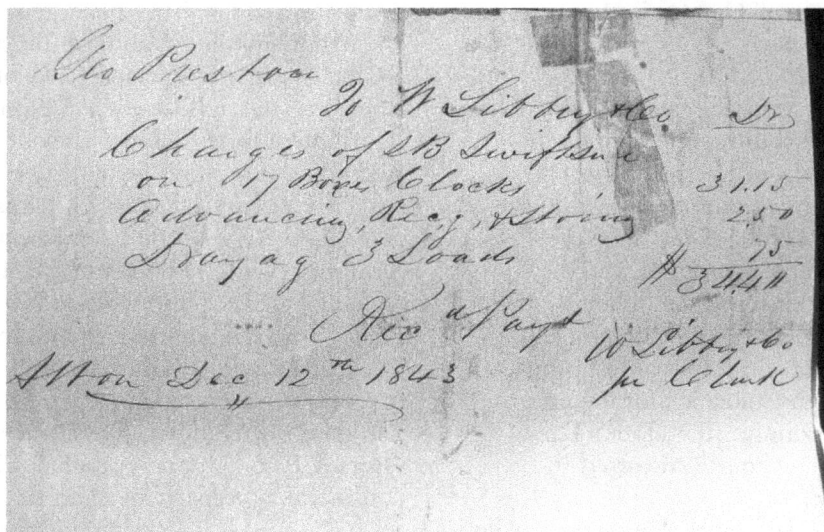

Figure 74. Receipt to Geo[rge] Preston, dated at Alton December 12, 1843, for shipment of 17 boxes of clocks. (Courtesy of Illinois Regional Archives Depository, Carbondale, IL.)

stopped at Brookville, IN, from whence he traveled to see "McKinny" on Williams, Orton, Prestons & Co. business. From there he went on to Alton, IL, where he boarded for several weeks at the home of clock factory worker and peddler Eliakim Thorp.[25] A receipt for taking clocks out of storage read: "Recd Cincinnati 1st August 1843 Mr. Geo Preston 21 Boxes Clocks in storage [signed] Geo Boggs Jr.,"[26] suggesting that the load of clocks referred to had been sent to meet Noah, traveling from Connecticut through Ohio via the Miami & Erie Canal. A receipt dated December 12, 1843, related to a second clock shipment, is shown as Figure 74.

Additional information about this second shipment indicated that the 17 boxes of clocks in it passed through Pittsburgh, suggesting the clocks had traveled across Pennsylvania by way of the Pennsylvania Main Line Canal, before being transshipped on the Ohio River, and then via the Mississippi River to Alton. W. Libbey & Co. of Alton, the forwarding and receiving firm that handled the shipment, stored some of the boxes of clocks in its warehouse.[27] From the above information it was also learned that Noah Preston's then 19-year-old son, George Marsh Preston, had been working as a clock peddler.

During the fall or winter of 1843 Noah traveled from Alton to Boonville, MO, on unspecified company business. In 1844, after paying the freight on a load of "clock parts," Noah went to Rockville IN, where he likely called on Ezra N. Marvin and George N. Marvin, both longtime clock customers of Williams, Orton, Prestons & Co., who had given the firm several promissory notes but had been slow to pay them. Returning to Illinois, Noah polished a batch of brass clock movements, purchased two boxes of glass, and then traveled to St. Louis to arrange for the purchase of a batch of clock weights.[28]

The probate file revealed a bill to Noah Preston from Henry C. Bull, requesting payment for boarding George Preston and George's horse from September 25, 1843, through May 30, 1844. At some point during 1843 or 1844, Noah and his son also visited the former Farmington, CT, clock maker George Marsh (the brother of Noah's deceased wife), in Ohio. During their visit Marsh lent the pair a total of $3. In June 1844, George M. Preston returned to Connecticut. He was back in Illinois by October 8, 1845, but returned to Connecticut soon afterward.

On July 14, 1844, one of Noah Preston's younger brothers, Gardner Preston of Harwinton, CT, wrote Noah a letter that began with the salutation, "Respected Brother." The letter—mostly concerned with personal matters—mentioned "George," "Sarah," and "Lucy," Noah's children, apparently all indirectly under Gardner's supervision, as doing well. Indeed, John B. Bartholomew billed Noah Preston for "Boarding your Daughter Lucy Preston from May 13th 1843 to December 1st 1844 20.59."[29] Bartholomew also boarded Noah's daughter Sarah Preston from May 1 to December 1, 1844, for which he charged Noah an additional $20.59.[30] As of the date of Gardner Preston's letter, Noah's daughters were quite young: Sarah Caroline Preston (born on June 19, 1837) was seven years old, and Lucy Adeline Preston, born on April 5, 1841, was only three. Gardner (sometimes spelled "Garner") had clock-making ties. He operated a small factory in Harwinton that produced 48,560 unpainted clock dials in the year 1845 alone.[31] Whether Gardner Preston produced clock dials for Williams, Orton, Prestons & Co. is unknown.

In June 1845 Noah Preston paid Robert Smith of Alton $8.50 for rent of the "clock shop" for eight and one-half months. He also visited a man named "Jacobs" on Williams, Orton, Prestons & Co. business. This was probably Charles C. Jacobs, who was indebted to Williams, Orton, Prestons & Co. for the hefty sum of $731 for ogee and gilded clocks Jacobs had purchased from the firm in the summer of 1838. Noah also paid for weight cord and a load of lumber in St. Louis. In addition to these activities, he shipped 12 pairs of clock weights to "Rock Island, Nauvoo, &c."[32] Both Rock Island and Nauvoo were Illinois towns on the banks of the Mississippi River about 200 miles north of Alton; Noah seems to have visited Nauvoo to sell a few clocks in the vicinity during his travels through Illinois and Indiana.[33] The mention of Nauvoo is particularly interesting because of its settlement in 1839 by members of the Church of the Latter Day Saints (Mormons), whose leader, Joseph Smith, was murdered by a mob on June 27, 1844, while under protective custody.[34] By 1846 Nauvoo, which had quickly become one of the largest towns in Illinois, was largely abandoned, many of its former Mormon residents moving farther west and eventually settling in Utah.[35]

Noah Preston happened to be working at Bunker Hill, Macoupin County, IL, a few miles northeast of Alton (see Figure 75) in 1845, when an accident occurred that not only altered the course of events for the firm of Williams, Orton, Prestons & Co. in Illinois but also ultimately ended Noah's life. On August 6,

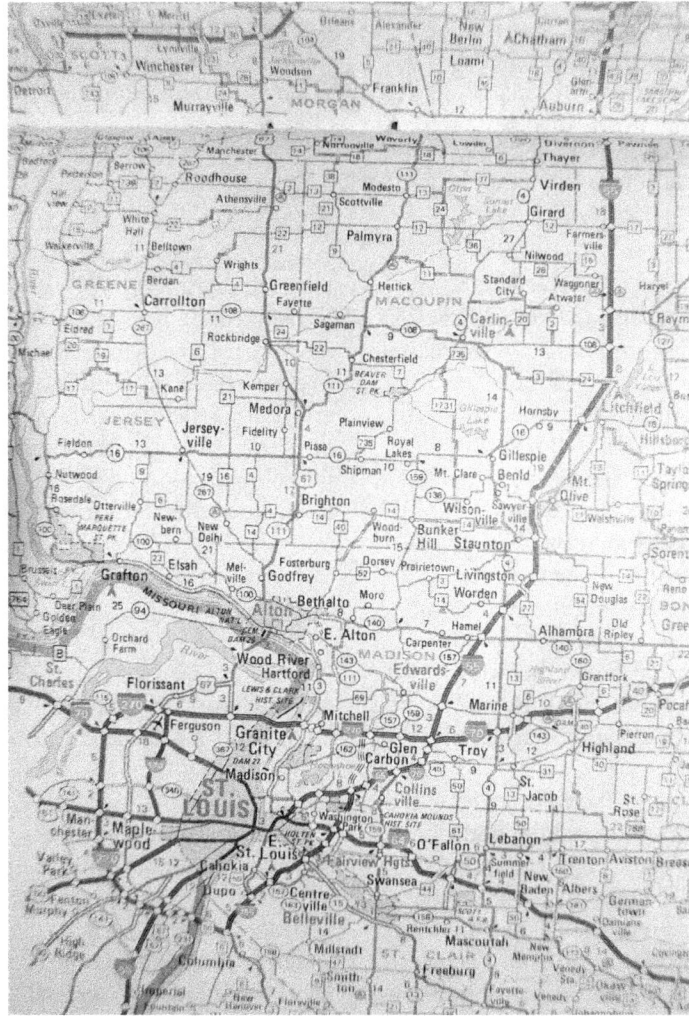

Figure 75. Portion of twentieth-century road atlas for Illinois, showing relative locations of Bunker Hill (lying north of Edwardsville), Alton, and Carlinville. (Source: AAA Road Atlas *(Falls Church, VA: AAA, 1987.)*

Figure 76. Notice of Noah Preston's death appearing in the Connecticut Courant, *September 5, 1845.*

Figure 77. Erastus Higby's bill, dated November 2, 1846, to Noah Preston's estate for nursing Noah during the latter's last illness. (Courtesy of Illinois Regional Archives Depository, Carbondale, IL.)

1845, he was severely injured after being run over by a loaded wagon.[36] Nearly two weeks later, on August 19, he died. On September 5, the *Connecticut Courant* notified its readers of the accident and of Noah's death (Figure 76). Much of the information about Williams, Orton, Prestons & Co.'s activities in Illinois that forms the basis for our findings, including the partnership agreement between Bishop, Woodford, and Williams, Orton, Prestons & Co. mentioned above, was gathered from the nearly 200 pages of documents comprising Noah Preston's Madison County, IL, probate file, obtained from the Illinois Regional Archives Depository, Southern Illinois University at Carbondale. Unless otherwise noted, the probate record was the source of the information presented below.

Bunker Hill resident J. A. Delano, probably the local druggist, was the first medical responder to arrive on the scene of the accident. He later billed Noah's estate $3.87 for medicines including "essence of peppermint", "soda powders", and for "Medical attendance", on several occasions between August 6 and August 20. Erastus Higby, still in Illinois, and still peddling clocks, arrived on August 7 to take charge of nursing Noah. Higby's bill to Noah's estate, dated November 2, 1846, is shown in Figure 77. Physician Frederick Humbert, a German immigrant who had settled in Upper Alton in 1836, arrived at Noah's bedside on August 7. Humbert made the 18-mile trip from Upper Alton two more times during Noah Preston's illness, on August 9 and 13, each time remaining with his patient overnight. On August 7, after consultation, Humbert inserted a "stomach tube" in Preston's "rectum and colon." On the night of the 9th, he dressed four of Noah's fractured ribs. Another physician, Dr. Ebenezer Howell, who had come to Bunker Hill in the spring of 1837, and for many years afterward was the town's only practicing physician,[37] arrived at Noah Preston's bedside on August 8. Thereafter, Howell made daily visits and provided unspecified medications until his last visit on August 16.

Despite receiving, as it were, the best medical care frontier Illinois could offer, as stated above, Noah Preston succumbed to his injuries on August 19. Nine days later, on August 28, Edward H. Davis of Bunker Hill billed Preston's estate for "Boarding N. Preston" one dollar per day for the 13 days of Noah's illness and also for "Bording Nurse" [i.e., Erastus Higby], and doing Noah's laundry, for a total of $20.

Despite the severity of his injuries, Noah managed to make a will (Figure 78). Dated August 8, 1845, it gave

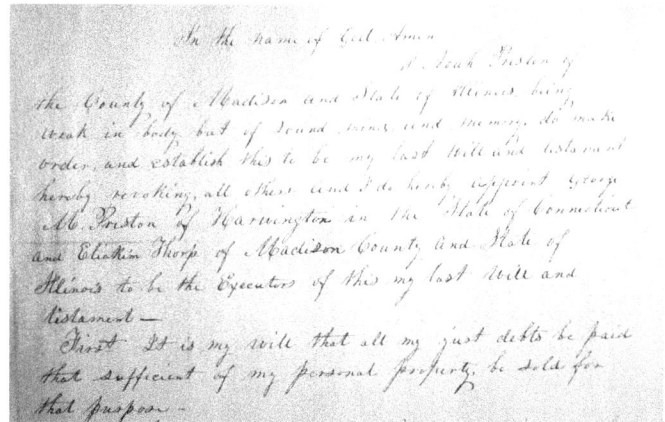

Figure 78. A portion of Noah Preston's will, dated August 8, 1845. (Courtesy of Illinois Regional Archives Depository, Carbondale, IL.)

Madison County, IL, as his place of residence. Noah named his son, George M. Preston of Harwinton, CT, and Eliakim Thorp of Madison County, IL, as the Executors of his estate. The latter was undoubtedly the same peddler with whom Noah had boarded upon his arrival in Illinois in 1843, and the same person Henry C. Bull had mentioned in his letter of July 13, 1840, to Williams, Orton, Prestons & Co. at Unionville. The relevant portion of Bull's letter (quoted in its entirety in Chapter 4) is presented again, below:

> ...[It] is of no use to hire men as long as the times Remain as they are. But Thorp thinks that he shall stay on the strength of the letter that you wrote last in which you gave it as the opinion of the Comp.[any] that they thought best for him to stay & work in the shop peddle & Collect & take Care of the Books But I have told him that we did not want him[.] But he thinks that the opinion of four [i.e., Cornelius R. Williams, Heman Orton, Eli Dewey Preston, and Noah Preston] is Better than one & that he shall stay for the Comp. now[.] I have no Objection to hiring Thorp if we wanted any one & should prefer him if he will work as Cheap as others but I am opposed to hiring him or any one at the present time...

I had a settlement with Thorp when his time was out & we owed him $275, for which I gave him our Note & in the settlement he Claimed $50 extra for keeping the Books which he says that <u>Wms </u>& N. Preston agreed to give him or that you would pay him what he thought was right, but he Concluded to take $45 & if <u>Wms</u> & N. Prestons say that if there was no such agreement that he would not Exact it he has not spent any Extra time in keeping Books & I think that he has had an easy time of it & that at the price that he had he could afforde to have done the wrighting without Extra pay but if you say that it is right then so be it...[38]

This passage from Bull's letter suggests Noah Preston and his partners (other than Bull) were well acquainted with Thorp, who was then about 27 years old. Born in Connecticut about 1813, Eliakim Thorp had married Cornelia P. Sperry there on November 26, 1832. However, Cornelia died soon afterward, at the age of 23. Based on the information provided in Bull's letter, Thorp, a cabinetmaker and mechanic,[39] arrived in Illinois to work for Williams, Orton, Prestons & Co. in 1839 or earlier. In Macoupin County on September 9, 1841, Thorp married (2nd) Frances A. [probably Amanda] Redman.[40] Sometime between 1842 and 1850 he served as the secretary of the Upper Alton Presbyterian Church Sabbath School Society.[41]

Not only did Eliakim Thorp remain an employee of Williams, Orton, Prestons & Co.'s Illinois branch, he also continued to help keep the firm's books. A note signed by Noah Preston read: "Upper Alton July 21st 1845 Mr E Thorp let Mr Wm Eaton tack [sic, - take] one wood clock if he calls it for the purpose of settling a note with Mr Abel Moore."

Noah's will specified that after paying his just debts, monies due him were to be collected and all his personal property sold. He bequeathed all the remaining monies to his son George, specifying that half of it was to be divided equally between George's two sisters, Sarah Caroline and Lucy, to be used for their support. When they arrived at age 18, the balance of their respective shares was to be paid to them. The remaining half of Noah's estate was to go to George.

Local attorney P. C. Huggins assisted Noah in preparing the will. Prior to 1847, Huggins, who served as a Macoupin County judge, was also Bunker Hill's only merchant.[42] His bill for services to Noah's estate reads as follows:

The Estate of Noah Preston

[1]845 In account with P.C. Huggins - Dr

Aug	To Writing Will	$1.00
	1 Pt Wine .20 1 do .20	.40
19	Shroud Shirt & Hkf 1.50 - 1 Shirt P. Board 5	1.55
"	Paying Pattrick for Digging Grave & Attendin[g]	<u>1.25</u>

[different handwriting:]
Recvd Payment $4.20

P.C. Huggins...[43]

George W. Carr and Martin Smith witnessed Noah's will. J. W. Cummins of Bunker Hill made a coffin for Noah's body on August 20.

The probate process on Noah's estate commenced on September 1, 1845, when Geo. W. Prichett, then Probate Justice of the Peace for Madison County, IL, summoned the witnesses to the will, George W. Carr and Martin Smith, to be brought before him at the Probate Office in Edwardsville, the county seat. The purpose of summoning the men was to require their testimony under oath about their presence (and perhaps Noah's condition) at the time the will was made. On September 2 the two subscribing witnesses were sworn in and signed the appropriate papers.

On September 30, 1845, Prichett certified that "a true copy" of Noah Preston's will was on "file and record" at his office. On the same day, a legal certification of Noah's death and a certification that the estate did not exceed $1,000 were both made.

By this time it was becoming clear that settling Noah Preston's estate, which was deeply entangled with that of Williams, Orton, Prestons & Co., would be a complex and time-consuming task. Consequently, both George M. Preston and Eliakim Thorp promptly declined serving as its executors. In their place the court appointed John A. Maxey of Madison County, IL, as the estate's administrator, with full power to secure and collect Noah's property and to pay his outstanding debts. On the same day, appraisers George Smith, Clark A. Moore, and Benjamin F. Long, after signing their oaths, were appointed and authorized to appraise Noah's estate. On September 30, John A. Maxey, with D. M. Kittinger (a local blacksmith and hardware dealer), and H. S. Summers (a local harness maker) as sureties, were bonded for $3,000 and signed the appropriate document.

At Upper Alton on November 11, 1845, the appraisers presented "A full and perfect inventory of the Goods and Chattels and personal estate of Noah Preston..." having a total appraised value of $1,016.35. Table 10 lists only the clocks and machinery in Noah's possession at the time of his death. On the basis of the number and assortment of clocks in the inventory, it is clear that at the time of his death, Noah was still involved in peddling them.

To the authors' surprise, the probate records revealed that Noah had been the copartner of one William Eaton Jr. of Upper Alton, IL, the inventor of "Eaton's Improved Threshing Machine," for which Eaton had received U.S. Patent No. 3,199 on July 28, 1843 (Figures 79A–79B).[44] The thresher was designed to clean and separate wheat "and other small grains" from their husks. The firm of Eaton & Preston had been established to manufacture and sell Eaton's Improved Threshing Machine. In addition to being Eaton's partner, Noah owned the exclusive right to use and vend the thresher in the State of Michigan during the continuance of Eaton's patent (see Table 10).

It appears Williams, Orton, Prestons & Co.'s Alton, IL, clock factory (or assembly plant) had recently been used for manufacturing activities other than clock making. A note found in Noah Preston's probate file reads: "Recd of Noah Preston A Note against John Seigh for twenty five dollars which when paid is in full for Rent of the building Called Clock Shop from the time the same was occupied by said Preston & Wm Eaton for building Threshing Machine up to the first day of May 1845[.] June 25th 1845 Robert Smith." So Noah rented the clock factory from its owner, Smith, up until May 1, 1845, for the purpose of building one or more threshing machines! Other fragments of paper revealed additional individuals involved in building the thresher. One reads: "Preston / to Enos Mc Cartney Dr / Threshing

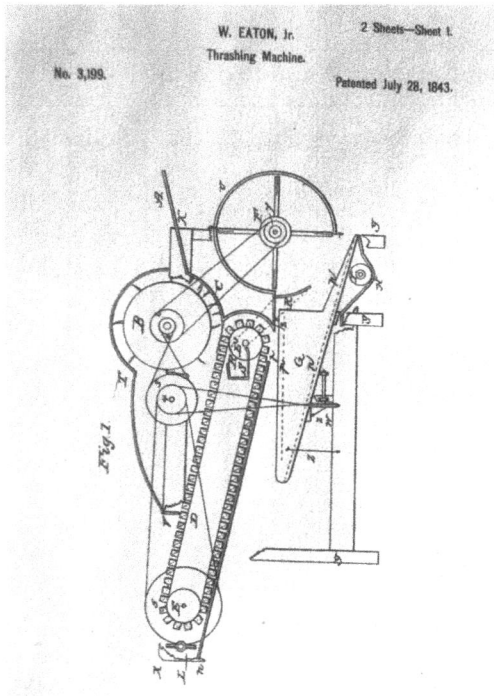

Figure 79A. Drawing (1 of 2) illustrating U.S. Patent No. 3199, July 28, 1843, issued to William Eaton, Jr. of Upper Alton, IL, for a threshing machine.

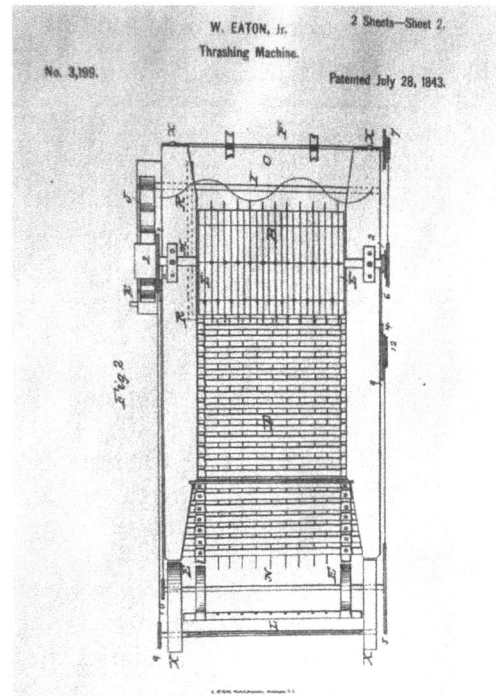

Figure 79B. Drawing (2 of 2) illustrating U.S. Patent No. 3199, July 28, 1843, issued to William Eaton, Jr. of Upper Alton, IL, for the threshing machine. (Figures 79A–79B courtesy of Google Patents [online source].)

Machine Cylinder 12.00 / Shell 1.00 / in work 1.32[?] / 1.6[?] / Threshing Machine 3.52[?] [total] 19[??]." A blacksmith's bill to the estate dating to September and October 1846 [1845?] included "Paid Baits for work on Machine," "G Preston," and individuals "S. Nutter," "Vallet," and "Green," apparently for various aspects of the work involved in producing one of the threshers.

In addition to the clocks and machinery mentioned above, Noah's estate contained items of a more personal nature: a "Sectional Map of Illinois," one Webster's Dictionary, a Psalm Book, a Bible, a German pipe, a "Rifle Gun," one "Smooth Bore [rifle] (broke)," three pairs of pants, one pair of boots "partly worn," an old coat, one "pigeons net," a carpet bag of clothes, and Noah's Newfoundland dog. There were also goods other than clocks clearly related to peddling, including 10 dozen gilt buttons plus another 10 10/12 dozen plain buttons, 5 pairs of children's shoes, 3 fur hats, 4 pairs of ladies' kid gloves, and a "lot knitting pins." Noah owned three horses and two sets of harnesses and harness trimmings. One "Bank Wheat Elevators" in the inventory may have had something to do with the threshing machine venture.

On November 17, 1845, all the above items were put up for sale by "crier" [auctioneer] Russ Hancock, with A. C. Robinson as clerk. John A. Maxey filed a report of the sale, which garnered a total of $1,100, at the Probate Office in Edwardsville on January 12, 1846. Several individuals recognizable for having been involved with Noah's business or his estate actively made purchases at the sale, including Henry C. Bull, Erastus Higby, Eliakim Thorp, Dr. Humbert, the constable who served the summonses on the witnesses to Noah's will, the estate's appraisers, and even John A. Maxey, the estate's administrator. However, none of the partners in Williams, Orton, Prestons & Co. other than Bull did so, so presumably neither they nor Noah's son George M. Preston were present.

After the sale the probate process on Noah's estate continued, with administrator Maxey filing a list of outstanding notes belonging to the estate at the Probate Office in Edwardsville on January 1, 1846. On March 7, the *Alton Telegraph & Democratic Review* announced that during "April next", the "Court of Probate of Madison County" would undertake the "final settlement of the estate of Noah Preston deceased" (Figure 80). But it was not to be the final settlement.

Table 10. Selected items from the inventory and sale lists of the estate of Noah Preston, deceased

Quantity	Item	Inventory Value	Selling Price
3	8-day Brass Clocks	@ $5 - $15.00	$17.38
2	130-hr. and 1 8-day wood clock*	@ $3 - 63.00	53.02
1	old silver watch	2.00	2.05
1	silver watch	6.00	4.25
1	Fancy silver watch	5.00	3.10
1	Endless chain	.50	.60
1	Cast Jack wheel & pinion	.50	.50
1	Eaton's improved Threshing Machine, Horse power, wagon, and tools belonging to Machine	250.00	296.50
1	Right of using and vending Eaton's Improved Threshing Machine in the State of Michigan during the fourteen years of the Patent right granted to Wm. Eaton	500.00	550.00
1	Brick Machine Frame & Castings	[not appraised?]	.25

*Two of the wood clocks had alarms.

Figure 80. Notice of "final settlement" on Noah Preston's estate appearing in the Alton Telegraph & Democratic Review, *March 7, 1846. (See text.)*

Sometime after the auction of Noah Preston's belongings in November 1845, Cornelius R. Williams arrived in Illinois to represent Williams, Orton, Prestons & Co.'s interests in the settlement of Noah's estate. While at Alton, it is said that Williams "...was prostrated with a severe sickness from the effects of which he never fully recovered."[45] Nonetheless, Williams lived for many years afterward, returning to Connecticut in 1847, settling in the section of Plymouth that became known as Terryville, where he died in 1880.[46]

Cornelius R. Williams had also come to Illinois to oversee the winding down of Williams, Orton, Prestons & Co.'s Illinois branch agency. On March 14, 1846, John A. Maxey appointed George Smith, W. N. Smith, and Eliakim Thorp to appraise the "Personal Property belonging to the late firm of Williams, Orton Prestons & Co." The total appraised value of the property was $517.69. On the same day as the appraisal, the property was sold at auction. This time, Cornelius R. Williams was among the buyers, purchasing six 30-hr. brass clocks, a "buggy wagon," and a lot of 50 clock-winding keys. Henry C. Bull was also a bidder at the sale, purchasing (among other items) a "lot of clock hands," a workbench, and a total of thirty-one 30-hr. wood clocks.

An example of a 30-hr. wooden movement shelf clock (Figures 81A–81C), with carved half-columns and a carved eagle splat, has a label that reads (in part): "Improved / CLOCKS / (With Brass Bushings) / MANUFACTURED AND SOLD / BY / Williams, Orton, Preston's & Co. / UPPER ALTON, ILLINOIS / AT WHOLESALE AND RETAIL..." (Figure 81B). Its familiar movement, Type 5.153 (Figure 81C), was made by Williams, Orton, Prestons & Co.[47]

Figure 81A. Exterior view, 30-hr. wooden movement shelf clock by "Williams, Orton, Preston's & Co." [sic], *Upper Alton, IL.*

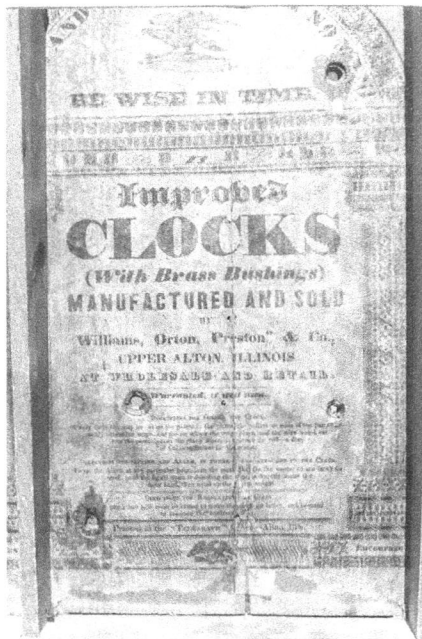

Figure 81B. Label of clock shown in Figure 81A.

Figure 81C. Williams, Orton, Prestons & Co. movement in clock (Type 5.153; see text), shown in Figures 81A and 81B. (Figures 81A–81C courtesy of Chris Brown, MJD photos.)

In addition to the items mentioned above, Henry C. Bull's purchases at the auction of Williams, Orton, Prestons & Co.'s property on March 14 included a total of ten 30-hr. brass clocks, 72 "Clock faces," a lot of "Winding keys," six "Steel bells," and other items, suggesting he planned to continue in the clock business. The sale netted a total of only $367.83, which was somewhat less than the property's appraised value.

Table 11 lists selected items sold at the March 14, 1846, auction of Williams, Orton, Prestons & Co.'s Illinois property. It has been estimated that the production of wood movements by all remaining makers ceased during the spring of 1847; Williams, Orton, Preston's & Co. had likely stopped manufacturing them several years earlier. The relatively low prices of about $2.50–$3.50 brought by the wood clocks in the sales of both the property of Noah Preston and that of Williams, Orton, Prestons & Co., support an argument that wood clocks were no longer profitable to produce.

The mention of "Lithographic prints" in the inventory is particularly interesting. Apparently, these were being used to decorate clock tablets.

In addition to the clocks, clock parts, and workbench mentioned above, included in the sale of Williams, Orton, Prestons & Co.'s property were various cabinetmaking tools, a "Chopping ax," a "Bake Oven," a chopping knife, a lot of knives, forks, and spoons, ten "Breakfast plates," nine "Dining" plates, a teapot, creamer, and sugar bowl, vegetable dishes, tin pans, stoneware jars and jugs, bedding, a desk, a bedstead and bed chord, and one "Brass Time piece." So the firm had provided living quarters for one and dining facilities for several individuals.

After the sale of Williams, Orton, Prestons & Co.'s personal property owned in Illinois came an important step in settling Noah Preston's estate, marked by a document transcribed as follows:

Table 11. Clocks, clock parts, and parts of clock cases, in the inventory and sales lists of Williams, Orton, Prestons & Co.'s Illinois property, March 1846.

Quantity	Item		Sold for
2	8 Day Ogee Brass clocks	@ 5.00	10.00
1	old 8 Day Wood clock		2.50
14	30 hour Coll Case Wood clocks	@ 3.50	49.00
58	30 hour Bronzed Wood clocks	@ 2.50	145.00
1	30 hour OG Wood clock		3.00
1	30 hour Alarm Wood clock		3.00
1	30 hour Old Wood clock		1.00
1	Brass Time piece		1.00
2	Old Brass clocks	@ 0.75	1.50
72	Clock Faces		2.50
5	Painted Glass		.25
19	Lithographic prints		1.00
Lot	Clock hands		.50
6	Wire Bells		.60
Lot	Bells mostly broken		.37½
Lot	pend Balls		1.50
Lot	Clock keys, hinge		2.50
Lot	parts of Cases		3.00
Lot	veneering		1.00
Lot	of Carving & frets		1.00
Lot	Sash hinges		.50
Lot	Brass Escutcheons		.50
1	Clock Screw fixing		1.00
21	Lithographic prints	@ 0.07	1.47

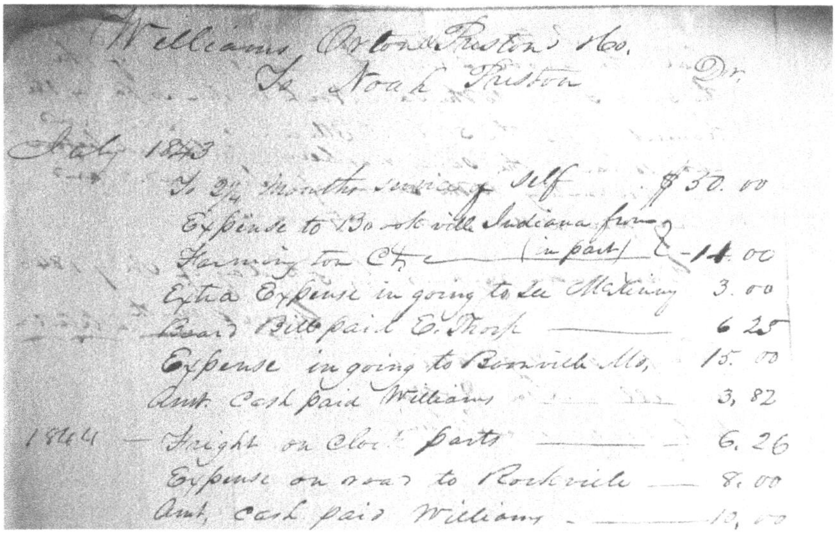

Figure 82. A portion of arbitrator George Smith's report dated May 5, 1846, showing Williams, Orton, Prestons & Co.'s outstanding debts to Noah Preston. (Courtesy of Illinois Regional Archives Depository, Carbondale, IL.)

We C.R. Williams and H.C. Bull Partners of the late firm of Williams Orton Prestons & Co. of the one part and John A. Maxey— Administrator of the Estate of Noah Preston deceased one of the Partners of the firm before named of the other part being anxious to settle and adjust the accounts of Noah Preston with said firm and the same being somewhat complicated do mutually agree to submit the same to arbitration of George Smith who is mutually chosen by us to examine and adjust all of said accounts, and to certify the same to the Judge of Probate… Witness our hands and seals this 28th day of April AD, 1846.

All three parties signed before a witness. On the same day, Maxey swore George Smith in as arbitrator. It may be recalled that Smith had been the first named appraiser in both the previous appraisals of Noah Preston's and Williams, Orton, Prestons & Co.'s estates, described above.

On May 5, 1846, the arbitrator returned his reports. In the first report, Smith listed the various expenses Noah Preston had incurred between July 1843 and June 1845 on behalf of Williams, Orton, Prestons & Co., for a total of $225.82. A portion of this document is shown in Figure 82. Noah's activities during this period

on behalf of the company have been described above. The largest expense, dated "July 1843," was "2 1/4 months service of Self $50.00". Beneath this report was a formal note, "To the Probate Justice for Madison Co.," from George Smith, arbitrator, stating: "[I] do certify that the above shows the Indebtedness of the said Firm to the said Noah Preston, so far as the transactions in the State of Illinois are concerned and so far as the same has been established before me…" Smith's second report showed that Noah Preston was indebted to Williams, Orton, Prestons & Co. in a total amount of $266.42, in part attributable to cash and notes belonging to the firm in Noah's possession at the time of his death. Noah also owed the firm for "19 10/12 doz. Buck Almanacks," cash received from Cornelius R. Williams "sundry times," and for "One Brass Clock," "12 Wood 30 hour Clocks," and "2 - 10 by 16 Looking Glass plates," the latter evidently to be put into clock cases. If the two debts were offset, it meant Noah's estate owed Williams, Orton, Prestons & Co. a net of $40.60. However, the matter was "laid over" until a "final settlement" of Noah's estate in Madison County Probate Court could be made.

Among the papers found in Noah Preston's probate file were barely legible copies of two notes that had been given to Williams, Orton, Prestons & Co. at Farmington, by a firm identified only as "Hill & Afflick," both "for value received." The first of the notes, in the amount of $1,653.25, was given during August 1838. The second, dated September 18, 1838, in the amount of $1,100, had been payable "on or by the 5th day of November 1838." However, it was not until July 18, 1839, that Hill & Afflick made a payment of $600 on the second note. Evidently, the remaining balances on both notes had gone unpaid. An undated account of sums owed to and by the firm of Williams, Orton, Prestons & Co., also found in the probate file, noted that judgment had been obtained against Hill & Afflick in a court case on the two notes.

Recalling that "Hill & Afflick" had also been mentioned in Henry C. Bull's letter to his partners in Williams, Orton, Prestons & Co. at Unionville, of July 13, 1840, the appearance of the notes from Hill & Afflick among Noah Preston's papers is interesting. A single shelf clock bearing the overpasted label of "Hill, Afflick & Co. / St. Louis, Mo.", undoubtedly the same firm, is known (Figures 83A–83C). Housed in a cornice-topped, modified triple-decker case (Figure 83A),

the clock contains a 30-hr. wood movement, brass-bushed, with the upper right-hand pillar post displaced to the right (Figure 83C). We have seen a very similar movement before, in a clock produced by George Marsh & Co. of Farmington village. The latter clock, label, and movement were shown in Figures 14A–14C of Chapter 1.[48] Nothing is known about the identities of Hill and Afflick, or the "& Co." in Hill, Afflick & Co., or what information might be found beneath the overpasted section of the clock's label.

On February 8, 1847, John A. Maxey presented an administrative account up to that date before the Probate Court at Edwardsville, showing that, with the proceeds of the estate sale of Noah's property, and after paying small debts, a total of $1,000 remaining in his hands was available for distribution to the estate's creditors. Agreeing, the Probate Court authorized Maxey to pay a "dividend" of 66⅔ cents on the dollar to a list of approved claimants relative to the amounts they were owed. At the same session of the Probate Court, Maxey presented an account of his own expenses on the

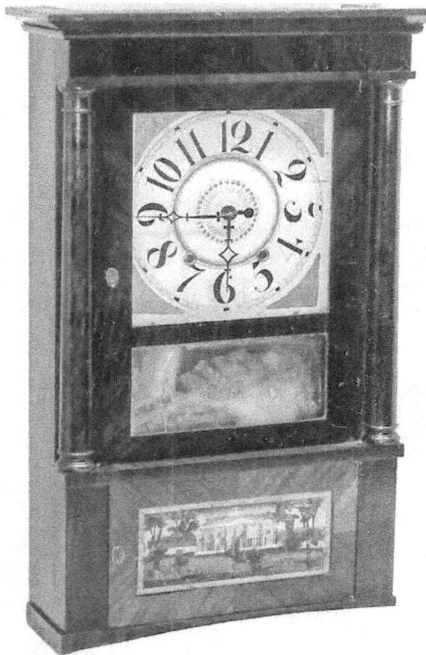

Figure 83A. Hill, Afflick & Co., St. Louis, MO, 30-hr. wood movement shelf clock in modified triple-decker case, exterior view.

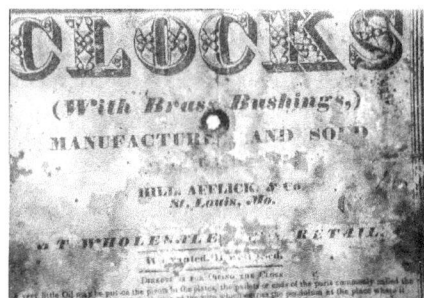

Figure 83B. Overpasted label of Hill, Afflick & Co., St. Louis, MO, in shelf clock shown in Figure 83A.

Figure 83C. 30-hr. wood movement in Hill, Afflick & Co., St. Louis, MO, shelf clock with overpasted label, shown in Figures 83A and 83B. Movement possibly by one of the Jerome firms of Bristol, CT, but similar to a few movements of George Marsh's firms or their successors--of which Williams, Orton, Prestons & Co. was one. (See text.) (Figures 83A–83C courtesy of Bryan Rogers.)

business of the estate, including expenses on a two-week trip to Rockville, IN, on unspecified matters, expenses in going "to Jersey county for property belonging to the estate," postages to Rockville and to Connecticut, a trip to Carlinville (the county seat of Macoupin County), and for several trips to the Probate Office at Edwardsville.

Recalling that through refinancing, Williams, Orton, Prestons & Co. (as well as partners Cornelius R. Williams, Eli D. Preston, and Noah Preston, individually) had remained solvent at the conclusion of the assignment of the firm's estate before Farmington District [CT] Probate Court on May 20, 1842, the existence of several outstanding lawsuits over unpaid debts in Connecticut illustrates the complex state of Noah Preston's finances at the time of his death. During the course of his efforts to gather and pay the estate's debts, administrator Maxey received information concerning the outstanding lawsuits. These are summarized in Table 6B as Court Cases LL through OO, Group V. In Case LL, Noah Preston had given Eunice J. Woodruff of Farmington a note dated March 1, 1838, in the amount of $87.50, on behalf of Williams, Orton, Prestons & Co. Eunice J. Woodruff seems to have been the daughter of Noadiah Woodruff of Farmington. However, the case records mentioned neither the goods or services she provided to the firm, nor that Eunice had died unmarried on September 25, 1844, at the age of 33,[49] before the note was completely repaid during the probate process on Noah's estate. Likewise, Noah's debts that were the subjects of Cases MM and NN were each settled in two installments.

It is unclear whether or how Case OO was ever settled as part of Noah's estate. This was the case of *Bryan E. Hooker vs. Noah Preston*, which commenced on July 28, 1843, about the time Noah left Connecticut for the West. Briefly, the case centered on Noah's assignment of the proceeds of a note in the amount of $350.33, to Bryan E. Hooker, the husband of Cornelius R. Williams's wife's sister, from the note's original holders, "George Marsh and Cornelius R. Williams, late partners and dealers under the firm of George Marsh & Co."[50] The note had gone unpaid. At the commencement of the suit, the sheriff's deputy attached property in Unionville described as "...bounded...North on land owned by Cornelius R. Williams and Eli D. Preston, East on land of Gideon Walker, South on land of Timothy Cowles, and West on land of Eli D. Preston and Cornelius R. Williams, and is the property of the...

Defendant [i.e., Noah Preston]."[51] However, on July 1, 1843, Noah Preston quitclaimed his interest in the former Williams, Orton, Prestons & Co. land and buildings to his brother, Eli D. Preston, and had also sold his nearby dwelling house to George Barber,[52] so it does not appear Noah had any remaining interest in the properties to attach. Subsequently, Noah defaulted on his appearance before the Hartford County [CT] Superior Court, and on October 13, 1843, the Court awarded execution in Hooker's favor. Nonetheless, as late as February 7, 1851, the Probate Justice of the Peace for Madison County, IL, H. K. Eaton, refused to authorize payment of the debt to Bryan E. Hooker, stating that the matter should have been settled by selling the property, and if not, the copies of court records sent from Connecticut should have explained why not.[53]

The remainder of the court cases summarized in Table 6B, Group VI, were initiated by others on behalf of Noah Preston or his estate, after Noah's death. More will be said about these cases shortly.

As mentioned previously, since his removal from Connecticut to Ohio in 1834, Noah's brother-in-law George Marsh had been busily accumulating property in the vicinity of Van Wert, OH, and elsewhere. Perhaps it is not surprising, therefore, that during the course of gathering Noah's property and paying his debts, administrator Maxey learned that Noah Preston and his son had also acquired land in Van Wert. An extant bill from G[eorge] Marsh debited Noah: "to paid taxes on lots in Van Wert Ohio owned by you & Geo M your son Time money &c $5.00." In lieu of currency, George Marsh took his pay in "two Clocks without weights." It is surprising, however, that the firm of Williams, Orton, Prestons & Co. also owned at least one parcel of land in Van Wert, which it held in the name of partner H.[eman] H. Orton, who is not otherwise known to have been a part owner of company land.

Aside from his Ohio land purchases, Noah Preston had also acquired property in Rockville, Parke Co., IN, near the Wabash River, and other property, apparently in connection with his partner in the threshing machine venture, William Eaton. However, additional issues related to Noah's financial dealings with Eaton needed sorting out. The subject of Court Case PP, Table 6B, *Cornelius R. Williams vs. William Eaton & Noah Preston*, was a note that remained outstanding and unpaid at the time of Noah's death, from the firm of Eaton & Preston to Williams, Orton, Prestons & Co. Williams won the case. Eaton appealed to an unidentified higher court in

Madison County but did not prevail, and on April 27, 1846, the Court awarded Williams $20.90 1/2 in damages plus another $10 in court costs. Subsequently, however, on June 25, 1846, William Eaton billed the $30.90 to Noah Preston's estate, adding: "To the frame Work of Thershing [sic] Machine as per agreement $17.5," less "Cr by work - done by [Eliakim] Thorp 5.83," giving "Balance Due $42.57." Although the nature of Noah's agreement with Eaton is not known, the amount was later paid from Noah's estate.

Several documents in Noah Preston's probate file pertain to Ezra N. Marvin and George N. Marvin (the latter possibly relatives), in Rockville, IN, each of whom was indebted to Williams, Orton, Prestons & Co. for a relatively large sum. The documents exemplify both the complexity of economic transactions in America caused by the lack of a national currency prior to the Civil War, and the state of Noah's finances. One of

---

Table 6B. Court cases brought by or against Noah Preston's estate.

The date of the writ and summons initiating the court case(s) is provided under "Date of Writ(s)." If the dispute was over non-payment or partial non-payment of a note, the amount of the note is given under "Amt. of Note"; the date of the note is listed in parenthesis below. A summary of the outcome of the cases, as far as known, is provided as "Outcome / Additional Info." Unless otherwise noted, the outcome was in favor of the plaintiff. Full references for each case are provided under "Reference." "Exn." = the date on which the court granted execution on its judgment (not necessarily the same as the date of the actual execution.) "Plff" = plaintiff; "Deft" = defendant; "Default" means the defendants failed to appear, in which case the court customarily rendered judgment in favor of the plaintiff; "Htfd. Co." = Hartford County; "Sup." = Superior; "CSL" = Connecticut State Library. All amounts are in dollars.

Case Date of Writ(s)	Title of Case	Amt. of Note / (Date of Note)	Outcome / (Additional Info)	Reference
Group V. Cases Outstanding at the Time of Noah Preston's Death				
LL 5/1/1841	*Eunice J. Woodruff v. Noah Preston.* X2	87.50 (3/1/1838)	Settled during probate process on NP's estate.	Farmington Justice of Peace Records, Box 41, RG62, CSL.
MM 5/1/1841	*Henry Mygatt, administrator on Noadiah Woodruff estate v. Noah Preston.*	228.60 (3/1/1838)	Settled during probate process on NP's estate.	Farmington Justice of Peace Records, Box 41, RG62, CSL.
NN 7/6/1843	*Timothy Cowles, executor of estate of Horace Cowles, for widow Elizabeth H. Cowles v. Noah Preston.*	23.42 (10/23/1840)	Settled during probate process on NP's estate. (Htfd. Co. Co. Court.)	Noah Preston Probate File Illinois Regional Archives Southern Illionis U., Carbondale, IL.
OO 7/28/1843	*Bryan E. Hooker v. Noah Preston.*	350.33 (4/8/1841)	Default; Exn. granted 10/13/1843, but never served. Unknown how settled.	Noah Preston Probate File, Illinois Regional Archive Depository, Southern IL U., Carbondale, IL.
Group VI. Cases Involving Noah Preston's Estate				
PP 3/28/1846	*Cornelius R. Williams vs. William Eaton & Noah Preston.*	20.57 (due 1/8/1845)	Judgment in favor of plff,; deft. failed on appeal in Madison Co. court system; Exn. in favor of Williams 4/27/1846.	Noah Preston Probate File, Illinois Regional Archives Depository, Southern IL U., Carbondale, IL
QQ (Waived)	*Thomas Nelson for John A. Maxey, admin. on Noah Preston's estate, v. Ezra N. Marvin.*	16.00 (Unknown) made on 8/13/1847 in an unidentified Parke	Plff to recover of deft 10.00 by agreement Depository, Southern IL U., Carbondale, IL.	Noah Preston Probate File, Illinois Regional Archives Co., Indiana court.

them is quoted as follows: "Received of Ezra N. Marvin a Note of hand signed by / Joseph Curtis for 51 dollars 64 cents to collect or return / Rockville May 21st 1844 [signed] Noah Preston / for Williams Orton Preston & Co." It was marked: "[a] true coppy of a receipt / held against Noah Preston / word for word [signed] Ezra N. Marvin." Another document reads: "Rockville January 10th 1845 / Mr Ezra N Marvin / Will pay Joseph Strain / Or beare[r] fifty dollars Monies I have / this day receipt from Said Strain / on your accounts and Oblidge / Yours &c [signed] Noah Preston." At Rockville on October 14, 1846, John A. Maxey took two notes on third parties, plus $27.30 in cash, from Ezra N. Marvin in payment on a "large note" due Noah's estate. However, it appears Maxey did not attend court in Case QQ (Table 6B) in person, when, waiving the customary writ and attachment, the Court offset the mutual indebtedness from Ezra N. Marvin and Noah Preston to one another, leaving a net balance of $10 due in favor of Noah's estate.

An additional bit of evidence related to Noah's dealings in Indiana, in the form of a notification, read: "Re'd [Received] in Store for N Preston / four boxes Tobacco to be Shipped / to Evansville first opportunity / Montezuma March 21st 1845 / [signed] Hughes & Adams / Per P Comwell." Both Evansville and Montezuma were (and are) in Parke County, IN. Montezuma, situated on the east bank of the Wabash River, served as a port for nearby Rockville. The four boxes of tobacco may have been the ones sold on November 17, 1845, at the sale of Noah's personal estate.

On October 9, 1849, John A. Maxey addressed the following letter at Upper Alton to Probate Justice of the Peace H. K. Eaton at Edwardsville:

> Dear Sir In Examining the business concerning Prestons funds – please to Say to me in your letter of Instructions if I may, Can or Shal[l] place a head and foot stone at his grave The pro[b]ible cost will be 10 or 15 dollars his frien[ds] at Bunker Hill request it to be done please write me fully on the business at large and oblige yours Respectfully [signed] J.A. Maxey [sic].

It is unclear whether Probate Justice Eaton subsequently authorized an expenditure for the placement of monuments at Noah's grave.

Headed very informally, "Prob off Ed. [Probate Office Edwardsville] March 30, 1848.," and signed informally "E." [H. K. Eaton] was a summary report, which reads, in part:

> Est Noah Preston dec^d And now at this day comes the Adm^r of Estate of N.P. dec^d [Maxey] and prays that he may be authorized to make payment in full of the balances due on claims against said Estate on which a partial payment was authorized to be made on the 8th of Feb. 1847. And the court having considered the matter, from the representations made, and it appearing that sundry large claims from Connecticut against said Estate on file in this office are not sufficiently authenticated and the claimants thereof having been written to by said Adm^r as the court has been informed by him, and they having failed to produce proof and also failed to appear & settle with said Adm^r and as said Adm^r says he has reason to believe that some of said claims have been paid; and that a fair settlement it would so appear in regard to some of them, if not all, and all at least would be subject to reduction, and as injustice is done to those creditors in this State whose claims have been duly proved and on file in this office; and said Creditors are unwilling to wait longer, this court from all these representations, concludes that it is consistent with justice to authorize said Adm^r to make full payment of said claims referred to...

Evidently, the "sundry large claims from Connecticut...not sufficiently authenticated" included those of Bryan E. Hooker (described above), and (as we shall see), a $500 claim made by the late firm of George Marsh & Co. There followed a list of eight names, mostly familiar, consisting of: Henry C. Bull, Almira Cummings of Bristol and Farmington, CT, Robert Smith, William Eaton, Eli D. Preston, Illinois blacksmith and tool dealer D. M. Kittinger, Illinois harness maker H. S. Summers, and Josiah Little, to whom Noah's estate was indebted for stabling horses in Illinois. All of these individuals were owed relatively small balances after receiving the partial payments from the estate that had been authorized in February 1847.

The final version of this document, signed "H.K. Eaton P.J.P.," did not mention the poorly substantiated

Connecticut claims. Otherwise, the list of claimants and the amounts owed them was identical, and a provision was inserted ordering receipts to be taken when the claimants were paid. The signed receipts, dated over the next several months, were among the documents found in Noah Preston's probate file.

Still working to settle the estate at Upper Alton on April 25, 1857, John A. Maxey sent "George Marsh & Co. by C.R. Williams Agt" $300.00. Similarly, on May 24, 1859, Maxey turned over another $200.00. Cumulatively, the two payments totaled $500.00. This was the return of the money that George Marsh loaned to Noah Preston on April 21, 1834, when Marsh went to Ohio and Noah, in Connecticut, formed Orton, Preston & Co.! On the 1859 transmittal document, Cornelius R. Williams's address was given as "Terryville / Connecticut." Williams wrote back asking for an accounting of clocks and other items in the inventory of Noah's estate, as if still concerned that they had belonged in part to the former companies. Nonetheless, he expressed his willingness to sign whatever "form of such writing as you [presumably Maxey] wish." Clearly, Williams was looking forward to closure in the matter.

In the end it took more than 20 years to settle Noah's estate. John A. Maxey, still its administrator on November 3, 1866, made a final list of his own expenses and also a final accounting on the estate. Briefly, the "amount of the estate" brought forward from Maxey's previous report of February 8, 1847, was $1,263.59. Additions from that time totaled $218.64, giving a new total of $1,482.23. The updated calculation of the outlays, including the debts paid in installments pursuant to the partial settlements of February 1847 and March 1848, and the large $500 payment to George Marsh & Co. made in two parts during 1857 and 1859, all totaled $1,056.38, which, when deducted from the $1,482.23, left the estate with a positive net balance of $425.85.

In his final accounting Maxey noted, "…a sale note on J.L. Bingham and H.A. Tappan of $550 which was…uncollected & was placed in the hands of Palmer & Petman for collection…" J. L. Bingham and H. A. Tappan had been the purchasers of Noah Preston's right to manufacture and vend Eaton's Improved Threshing Machine in the State of Michigan for $550 at the auction sale of Noah's property on November 17, 1845. Bingham and Tappan had given the estate their promissory note but had failed to pay it. Consequently, about 1853, Maxey brought suit against them in Macoupin County Circuit Court. During the trial, Bingham and Tappan argued that Eaton's improvements had been made by others, and that the thresher was inferior to others then in use, thus basing their claim that the "consideration of the note...[had] wholly failed."[54] When the Circuit Court decided in favor of the plaintiff (i.e., Maxey), Bingham and Tappan appealed to the Illinois Supreme Court, which upheld

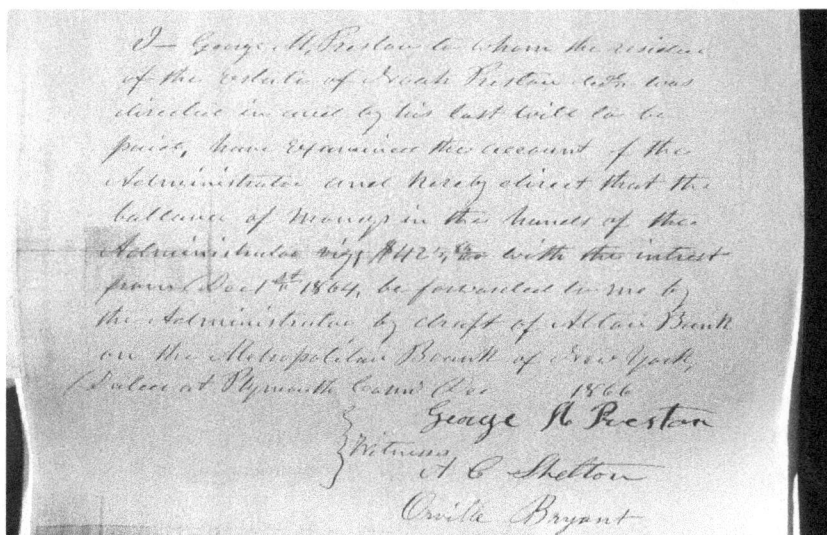

Figure 84. George M. Preston's acceptance of administrator Maxey's final account of Noah Preston's estate, December 1866. (Courtesy of Illinois Regional Archives Depository, Carbondale, IL.)

the Circuit Court's ruling. The final words in the report of the case of *J. Bingham et al. vs. Maxey* plaintiffs and defendants in error, respectively, argued before the Illinois Supreme Court's 1853–54 terms, read: "In the absence of fraud, caveat emptor."[55] Subsequently, according to Maxey's 1866 report, the amount was collected.

In addition to the suit against Bingham and Tappan, Maxey apparently initiated a suit against George N. Marvin of Rockville, IN. Near the end of Maxey's final report was noted: "On the 2 notes of George N Marvin, on trial it was shown that $163.44 only was due [to Preston's estate] and [was included in] Administrator accounts…in his former settlement."

Just above Maxey's signature on his final report of 1866, he noted: "All the notes uncollected & accounted for have been deposited and filed for the benefit of those concerned." These notes, amounting to the sum of only $36, remained uncollected because of the insolvencies of the respective debtors. Therefore, it seems that John A. Maxey, administrator on Noah Preston's estate, toiling from 1845 to 1866, had done a good job.

Finally, on December 27, 1866, George M. Preston, then of Plymouth, CT, indicated his acceptance of Maxey's final account of his father's estate in writing (Figure 84). The document states:

> I – Geroge M. Preston to whom the residue of the Estate of Noah Preston dec[d] was directed in and by his last will to be paid, have Examined the account of the Administrator and hereby direct that the balance of Moneys in the hands of the Administrator viz $425, 85/100 with the interest from Dec 1st 1864, be forwarded to me by the Administrator by draft of Alton Bank on the Metropolitan Bank of New York...

Below George M. Preston's signature was noted: "First National Bank of Alton Draft No 4239 on Metropolitan Nat Bank New York date Jany 4 1867 to order of Jno A Maxey Esqr for $425.85 Endorsed by Jno A Maxey to order of Geo M. Preston sent by mail to Geo M Preston Plymouth Con Jany 21 1867 [signed] C.W. Dimmock / Clerk." So ends the commingled stories of Noah Preston and the former firm of Williams, Orton, Prestons & Co. of Farmington, CT, and Alton, IL.

In Chapter 6 we will return to Connecticut to examine the decline of wooden movement clock making in Unionville village after the dissolution of Williams, Orton, Prestons & Co. In so doing, we will attempt to cast light on the careers of some of Unionville's most obscure clockmakers.

Notes and References

1. Kenneth D. Roberts and Snowden Taylor. *Eli Terry and the Connecticut shelf clock.* 2nd ed. Fitzwilliam, NH: Ken Roberts Publishing Co., 1994.

2. Julius Gay. *An historical address delivered at the annual meeting of the Village Library Company of Farmington, Connecticut, September 9, 1903.* Hartford, CT: Case, Lockwood, Brainard Co., 1903.

3. W. T. Norton, N. G. Flagg, J. S. Joerner, eds. and compilers. "History of Alton Township, Madison County, Illinois." In: *Centennial history of Madison County Illinois and its people.* Chicago, IL: Lewis Publishing Co., 1912.

4. "Research Activities & News." NAWCC *Bulletin*, No. 264 (February 1990): 76–78.

5. Ibid.

6. Ibid.

7. According to W. T. Norton, N. G. Flagg, and J. S. Joerner in their "History of Alton Township, Madison County, Illinois," the city today known as Alton was begun about 1814–17 in several locations. These included two called "Alton," one called "Upper Alton," another called "Alton on the River," and a village situated north of Upper Alton known as Salu. However, because most of the settlers came from New England, the various Altons were referred to jointly as "Yankee all town."

8. James H. Whitaker. "Williams, Orton, Prestons & Co., Upper Alton, Illinois." NAWCC *Bulletin*, No. 268 (October 1990): 460–461.

9. Mary Frances MacKie. *Avon, Connecticut, an historical story.* Canaan, NH: Phoenix Publishing, 1990.

10. Noah Preston Probate File, Illinois Regional Archives Depository, Carbondale, IL.

11. Frank B. Woodford. *The Woodford family record.* Detroit, MI. 1938: 278–280.

12. *Alton Marine & Fire Insurance Co. v. John Manning et al.* Clerk of the Circuit Court, Madison County, Edwardsville, Illinois.

13. Ibid.

14. Rochelle M. Joseph to MJD, January 16, 2009. The names provided to the archive were those found in the foreclosure notice. See also: *The people [of the State of Illinois] ex rel. Fishbourne v. Higby*, Marshall County Circuit Court [files], Illiniois Regional Archives Depository, University of Illinois, Springfield, IL.

15. *The people [of the State of Illinois] ex rel. Fishbourne v. Higby*.

16. Ibid.

17. Lookingforlincoln.com (online source).

18. *The people [of the State of Illinois] ex rel. Fishbourne v. Higby*.

19. Ibid.

20. Ibid.

21. Dan W. Bannister. *Lincoln and the common law: a collection of Illinois Supreme Court Cases from 1838-1861 and their influence on the evolution of Illinois common law.* Springfield, IL: Human Services Press, 1992: 3.

22. Mary Frances MacKie, *Avon, Connecticut*, 1990.

23. H[enry] C. Bull to Williams, Orton, Prestons & Co., July 13, 1840, private collection.

24. *Austin F. Williams vs. Joseph Bishop et al.* Town of Farmington, Justice of the Peace Records 1840–1859, RG62, Connecticut State Library.

25. Noah Preston Probate File, Illinois Depository.

26. Ibid. Based on primarily source data for other Connecticut clock making firms, each box may have contained as many as five or six clocks.

27. Ibid.

28. Ibid.

29. Ibid.

30. Ibid.

31. "Statistics of the Town of Harwinton for the year ending October 1st, 1845." Hungerford Library/Museum, Harwinton, CT.

32. Noah Preston Probate File, Illinois Depository.

33. Undated, unidentified accounts in the Noah Preston Probate File, apparently in Noah's handwriting, but with notations made by other persons, indicate that Noah visited several towns in Illinois and Indiana in addition to those mentioned, between the time he left Connecticut in 1843 and the time of his death on August 19, 1845.

34. Fawn M. Browdie. *No man knows my history: the life of Joseph Smith*. New York, NY: Alfred A. Knopf, 1963.

35. Harry E. Pratt, ed. "Illinois as Lincoln knew it." In: *Papers in Illinois history.* Springfield, IL: Illinois State Historical Society, 1938: 109–187.

36. *Hartford Daily Courant*, September 5, 1845.

37. Charles A. Walker, ed. *History of Macoupin County, Illinois.* Chicago, IL: S. J. Clarke Publishing Co., 1911. See also: history.rays-place.com (online source); and www.bunkerhillhistory.org (online source).

38. H[enry] C. Bull to Williams, Orton, Prestons & Co., July 13, 1840.

39. Connecticut Vital Records, Church, Marriage, and Death indices, Connecticut State Library. See also: 1850 U.S. Census, Upper Alton, Illinois.

40. Illinois Marriages 1790–1860. Ancestry.com (online source).

41. William L. Clements Library, University of Michigan, Manuscripts Division Finding Aids. "Upper Alton [IL] Presbyterian Sabbath School Society Minutes 1842–1850." quod.lib.umich.edu (online source).

42. Charles A. Walker, ed. *History of Macoupin County, Illinois.*

43. Noah Preston Probate File, Illinois Depository.

44. Google Patents (online source).

45. Clarence R. Williams, compiler. *Washington Williams of Rocky Hill, Connecticut.* (Unpublished manuscript, 1943). Connecticut State Library.

46. Ibid.

47. Snowden Taylor. "Update on Terry-type 30-hr. wood movements." NAWCC *Bulletin*, No. 249 (August 1987): 314–315. A further update to renumber this movement as Type 7.153 is proposed (Snowden Taylor, unpublished).

48. See: Snowden Taylor. "Research Activities & News." NAWCC *Bulletin*, No. 281 (December 1992): 731. Some confusion surrounds the present classification of the movement of the Hill, Afflick & Co. clock shown in Figure 83C. In an unpublished update to his 30-hr. wood movement classification scheme, author ST proposes renumbering it as

Type 7.113, with one of the Jerome firms as its maker. However, this movement is quite similar to the proposed Type 7.116, 30-hr. wood movement made by George Marsh's Farmington firms (or their successors), discussed in Chapter 1, shown previously in Figure 14C. The displaced upper-right pillar post is a feature of both movements. An obvious difference between the two movements in question appears to be the large brass winding arbor bushings on the Hill & Afflick example.

49. Susan W. Abbott, compiler. *Woodruff genealogy.* (Milford, CT, 1963.)

50. See Chapter 1 for discussion of the firm of George Marsh & Co. of Farmington village.

51. Noah Preston Probate File, Illinois Depository.

52. Farmington Land Records (FLR) Volume (V.) 45, p. 461, and V. 48, p. 433, July 1, 1843.

53. Noah Preston deeded the property to George Barber on July 1, 1843, shortly before Hooker brought the suit on July 28 (see FLR V. 48, p. 433).

54. E. Peck. *Reports of cases determined in the Supreme Court of the State of Illinois Nov. term 1853–June term 1854, inclusive.* Vol. XV. St. Louis, MO: W. J. Gilbert, 1870.

55. Ibid.

Chapter 6

"Minor" Clockmakers and Firms of Unionville Village: Virgil C. Goodwin; Goodwin & Frisbie; and Goodwin & Humphrey

Despite intense competition from increasing numbers of individuals and firms going into the clock business, the demand for 30-hr. wooden movement shelf clocks was still on the rise in 1832, the year in which a little-known clockmaker from Plymouth, CT, Virgil C. Goodwin, reached the age of 19. At that time the young United States of America was experiencing a speculative boom. In contrast, by 1839, the year in which shelf clocks bearing Goodwin's own labels originating at Unionville village began to appear,[1] the nation's economy was deeply mired in depression, with no real end in sight. While the recent introduction at nearby Bristol of 30-hr. shelf clocks with movements stamped from rolled sheet brass proved encouraging, less and less profit was being squeezed from shelf clocks with 30-hr. wooden movements, and Connecticut's larger clock-making firms began phasing them out. Thus, the wooden clock industry, which had enjoyed wild success over the previous decades, began its inevitable decline.

Precisely how old Virgil C. Goodwin was when he entered the trade that came to occupy much of his professional life is unknown. However, prior to 1855, children as young as the age of nine labored in Connecticut's manufacturing workforce,[2] and it is likely that Goodwin had been a youthful employee of one of the pioneering wooden movement clock factories in his native town of Plymouth. Indeed, as we shall see, although he may be considered a "minor" clockmaker in the relatively small numbers of clocks in which either his own, or one of his firms' labels appear, Goodwin's background and his relationships with other clockmakers of Unionville village and beyond, place him at the center of the present chapter in our chronicle.

Virgil C. Goodwin

Virgil C. Goodwin (1813–77) was born at Plymouth, CT, the sixth child and fourth son of Ozias and Asenath (Pond) Goodwin. On August 1, 1832, the day before his 19th birthday, Goodwin was familiar enough with the various aspects of clock making to enter into a written agreement with Henry Hart, a merchant of nearby Goshen, to produce 900 wooden shelf clock movements for Hart. A few weeks later, on September 13, 1832, the young Goodwin, evidently feeling able to support a family, married Fanny Olive Potter of Plymouth.[3] Before the movements could be completed, however, the agreement between the two men fell apart, and Goodwin initiated a lawsuit against Hart in the Litchfield County County Court. Details of the agreement and ensuing events found in the extant court case files[4] are presented below.

Henry Hart (1785–184?), the defendant in Goodwin's suit, was the second son born to David and Hannah (Hudson) Hart of Long Island, NY, and of Torrington and Goshen, CT. David Hart was a carpenter who trained several of his sons, including Henry, in various branches of woodworking. The Harts owned extensive properties in Goshen and Torrington, including a sawmill and clock factory straddling the boundaries between the two towns.[5] Hibberd's *History of the Town of Goshen, Connecticut* states: "...clocks [were manufactured] at Hart Hollow by Henry Hart."[6] Samuel Orcutt, in his *History of Torrington, Connecticut*, also mentioned: "...at a place called Hart's hollow, in the edge of the town of Goshen, quite a business was conducted in making clocks, about 1820 [sic; probably closer to 1830]; a number of buildings were erected and for a time the place assumed considerable importance..."[7]

A prominent member of the local community, Henry Hart was 47 years old when he entered into the contract with Virgil C. Goodwin. In addition to owning several sizable parcels of land in Goshen and Torrington, both within and beyond the boundaries of Hart's Hollow,[8] Hart served as a deacon of Torrington's Congregational Church and had represented Goshen at the Connecticut State legislature in 1822.

Consistent with evidence provided in the records of Goodwin's suit, sometime prior to October 1834, Henry Hart left Connecticut, heading to the American West where he intended to sell clocks then being made at Goshen. Taking up residence in the vicinity of Farmington, Fulton County, IL,[9] Henry Hart died in Illinois sometime prior to 1850.[10]

A surviving account book in the collection of the American Clock & Watch Museum of Bristol, CT, documents the clock-making activities of Alpha Hart, one of Henry Hart's brothers, between 1830 and 1834. A previously published analysis of the Alpha Hart accounts by author ST explained: "Alpha's brother Henry was, or became, involved in the clock business in his own right. Clocks have been seen with labels of 'Henry Hart' [see, for example, Figure 85, in which the firm of Jeromes & Darrow actually produced the clock for Hart[11]], and also 'Henry Hart & Son'. The former often have Jerome movements. The latter have various movements..."[12] So in hiring the young Virgil C. Goodwin, Henry Hart, who had often purchased movements for casing at his family's facilities, and even entire clocks to market under his own labels, evidently decided to have a batch of his own made on-site.

Virgil C. Goodwin's suit against Hart commenced on October 24, 1834, when a justice of the peace commanded the sheriff of Litchfield County to attach:

:..the Goods or estate of Henry Hart of said Goshen to the value of Eight Hundred Dollars - and for want thereof attach his Body and him have to appear before the County Court, to be holden at Litchfield within & for the County of Litchfield, on the third Tuesday of December A.D. 1834 then and there to answer unto Virgil C. Goodwin of Goshen aforesaid in a plea that to the Plaintiff the Defendant render the sum of Six Hundred Dollars, which to the Plaintiff the Defendant justly owes by Book...[13]

On the same day, a constable reported the attachment of the defendant's property, including "...two hundred Clock Cases, and Sashes, one wheel engine and apparatus, about One thousand parts of Clock movements One lot of iron Wire twenty five Clock movements, one Stove & pipe, a quantity of mahogany, and a lot lumber..."[14]

Attaching this much property was a labor-intensive undertaking, as revealed in the following tally of the constable's fees and charges:

P[ai]d. M. Goodwin for moving Clocks
with 3 hands & Team.---------------}   6.00
To myself & 2 hands with Team
one day moving Clocks-------------}   3.00
                                                    $9.96[15]

Figure 85. Label of clock made by the Bristol firm of Jeromes' & Darrow for Henry Hart of Goshen. (Courtesy of Bill Willard.)

Whether "M. Goodwin" might have been a relative of Virgil C. Goodwin's is unknown.

During the Court's September 1835 term, the defendant (Hart), by his attorney, Tho[mas] Miner, pleaded: "...that he owes the Plaintiff nothing in manner & form as the plaintiff in his Declaration hath alleged and of this puts himself on the jury for trial...."[16]

The case continued to the December 1835 term of the Litchfield County County Court, when it was put before a jury. Shortly thereafter, the jury returned its verdict as follows: "...vis: In this Case the jury find that the Defendant owes the Plaintiff nothing, in manner and form as the Plaintiff in his Declaration hath alleged - and therefore find for the Defendant to recover of the Plaintiff his Costs...."[17]

But Goodwin promptly appealed the verdict to the Litchfield County Superior Court, which began hearing his appeal during its February 1836 term. The case continued during the August 1836 and February 1837 terms, before the court rendered its decision in August 1837. We will examine the case's outcome momentarily.

The records of the trial of the case reveal much information of interest. The Court appointed Ansel Sterling, Thomas Mitchell, and David C. Sanford auditors to review the plaintiff's and defendant's books. The auditors' report, "respectfully submitted" on July 19, 1837, described their meeting with the parties and their respective attorneys, and their examination of the parties' respective claims, as follows:

> ...the said parties appeared before us with their books- and accounts and on that and the next succeeding day were fully heard...And thereupon- we find that the Defendant is in arrears and does owe to the Plff by Book to ballance book accounts between them the sum of four hundred Sixty four dollars & fifty cents of principal and the sum of Seventy four dollars of interest making the sum of $538.50 in the whole- & principal - which sums we award to the Plff- to be recovered of the defendant in this action accordingly...[18]

The matter did not end there. During the August 1837 session of Litchfield County Superior Court, Henry Hart, through his counsel, objected to the auditors' report. The court's record of his objections, quoted in part as follows, sheds light not only on the contradictory evidence found by the auditors, but also on relevant information from a horological perspective:

Now the D[e]f[endan]t in Court objects and remonstrates against the report made and presented by the auditors appointed in this case because he says that a part and most of the amount exhibited on oyer[19] and presented to and acted & Decided upon by said auditors consisted of an item and charge for work and Labor Done and performed by the Pl[ainti]ff on the running parts or movements of nine hundred & fifty clocks at the sum of Seven hundred Dollars- to which item and charge in evidence on the trial before said auditors the dft objected and also objected to the admission of the Plff as a witness to testify thereto, on the ground that said work and Labor (if performed at all) was Done and performed under and in performance of a Certain written agreement made and entered into this 1st day of August AD 1832 between Henry Hart on the one part & Virgil Goodwin on the other part witnesses that said Virgil hereby agrees to make clock movements of a good quality for said Hart and as many and as fast as said Hart may want for the term of [blank space] and to do all the wor[k] necessary to make them complete except the joiner work viz. dressing out wheel stuff plates clicks & bridges, excepting also the [blank] work, namely pendalum calps [sic,] crown wheels, bridges, buttons, studs, hands & springs and castings, namely wind keys, center tips, hammer. the said Virgil is to furnish his own board and light, and to make said clocks in a neat and workmanlike manner, being themselves [sic] responsible to said Hart for all damages if said clocks prove otherwise than well made and good the said Virgil farther agrees to make or procure all the tools, patterns and all the apparatus necessary for convenient making of s^d clocks, excepting wheel & pinion Engine, Lathes, Drill frames & filing burs & tools for curling wire, and to keep all tools for making said clocks in good repair, and said Hart agrees to furnish shop, fuel, and stock for tools and clocks, to prevent unnecessary delay in manufacturing said clocks and to pay said Virgil seventy five cents per clock for making and to make settlement yearly after the first five hundred

Henry Hart
Virgil C. Goodwin...[20]

Portions of Hart's "remonstrance" (objection) as they appear in the court case files are illustrated in Figures 86A and 86B. Note that the agreement between Hart and Goodwin excluded "joiner work" and also brass parts, hands, and "castings." The record went on to describe additional circumstances that led to the auditors' findings, as follows:

...on or about the 30th of August 1833 said written agreement was given up & abandoned and that a new verbal agreement was made between the said parties in all respects substantially like the written agreement before stated with the exception that the D[e]f[endan]t should pay or advance to the Plff from time to time during the progress of said work such sums of money as should be called for by the Dft [sic]

It was also claimed & found by the auditors that the Pl[ainti]ff under and in pursuance of sd agreement on or about the [space] Day of October 1834 had Done and performed most of the work and Labour necessary to complete said movements but that the parts of none of them then were or ever since have been put together and that no one of them was completed or finished-

It was also claimed by the Plff & so found by the auditors that a few weeks previous to the said [space] Day of October 1834 the Dft had left for the western country and so remained & was then absent. and that before he left he had appointed and constabled his son W[illia]m Hart his agent to facilitate the completion of sd Clocks & to forward them to him to the State of Illinois to be disposed of and that the said William remained in said Goshen for that it not being claimed or proved that the sd Wm had any other above mentioned authority therein. It was also by the plff further claimed & so found by the auditors that the plff on the day of October aforesaid requested the said William to furnish a small quantity of wire and a small number of [center tips?] which he claimed to be necessary to finish & complete said work but that the said William refused to furnish the same. and also that the sd William told the Plff that he need Do no more to said clocks- (but it was not expressly admitted by the Plff that he gave no notice to the said William that he assented thereto, or that he would or should not finish & complete said clocks or that the Plff on his part would stop and not complete said contract by finishing said movements.) and the Plff offered himself as a witness to have[?] that certain materials were needed to finish said work & that he called upon the said William to furnish the same and that he refused so to do- and also to prove that the said William told him the said Plff that he need do no more to sd clocks & that he avoid any further performance of sd agreement - to which testimony of the Plff the Dft objected but the auditors overruled sd objection and admitted the Plff so to testify which he accordingly did-

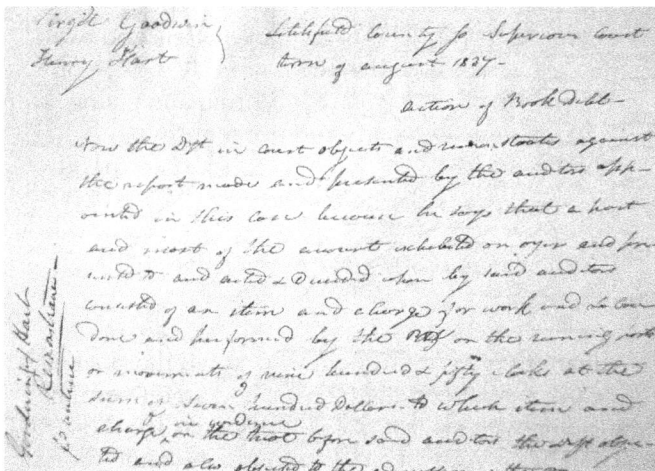

Figure 86A. Beginning portion of document containing Henry Hart's remonstrance before the August 1837 session of Litchfield County [CT] Superior Court in the case of Goodwin vs. Hart.

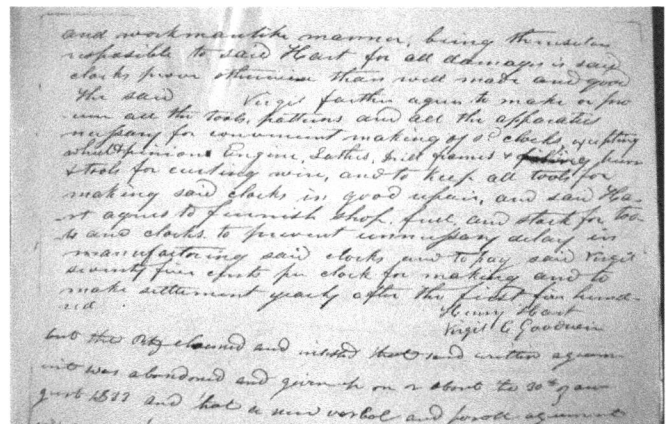

Figure 86B. Signature portion of Henry Hart's remonstrance shown in Figure 86A, describing his agreement of August 1, 1832 with Virgil C. Goodwin. (Figures 86A–86B, courtesy Connecticut State Library.)

It was also proved & found by the auditors that the Plff did not & had not completed said movements nor any of them, but that immediately after on the said Day of October he the Plff brought commenced [sic] this his action on Book by writ of attachment against the Dft, and by virtue thereof caused said clock movements to be attached & held in the custody of the Law where they ever since have & do now remain unfinished - it being admitted that the Dft had never accepted or received said movements or any part thereof, and that there was no agreement that he ever should or would receive or accept the same-

Whereupon the Dft [i.e., Hart] claimed and insisted before the auditors & does now insist that the action of Book debt will not be & ought not & cannot be sustained to recover for the work & Labor done & performed by the Plff upon said clock movements. and that sd account ought not to have been admitted in evidence nor the Plff admitted to testify thereto or to the sundry other facts herein before set forth. but the deft says that sd auditors did admit sd account in evidence, and did admit the plff to testify thereto & to all the facts above told and in the settlement of the accounts of the parties did allow for the work & labor on sd movements the sum of $700. [?] & a [?] sum as interest. and did also divide all the introductory questions as to the admission of evidence and c. as above stated all which he is ready to verify wherefore the deft says the court ought not to accept sd report but to reject the same.

By Ph Miner for the deft[21]

Despite Hart's protest, the Court accepted the auditors' report, finding that Hart owed Virgil C. Goodwin a total of $538.50 for the work he performed, plus Goodwin's sizable costs of suit, now amounting to $117.72. According to the record, "Ex[ecutio]ⁿ issued Augᵗ 30. 1837."

The file contains an interesting full-page list calculating the plaintiff's court costs, entitled: "Virgil C. Goodwin vs. Henry Hart, Dec. Term 1834, Plffts. Bill of Cost". Each expense was itemized, with total expenses of $117.72, as indicated above. Included on the list were the costs "for storage & keeping" of the clock movements, cases, and the wheel engine that had been in Henry Hart's clock factory at the beginning of the case, but which had been attached by the constable in October 1834, prior to the parties' first scheduled appearance in court. In addition, fees and expenses for the April and September 1835 terms of the county court included expenses for the jury, and for several witnesses, including Dan Goodwin, Ozias Goodwin, and George Killbourn [sic]. The same three witnesses were also called during the subsequent trial of the case on appeal. It appears that the Ozias Goodwin who testified in the case was Virgil Goodwin's older brother, who was born in 1809. Another of Virgil Goodwin's older brothers, Dan Harrison Goodwin (1811–96),[22] was undoubtedly the "Dan Goodwin" who was called as a witness. Witness George Kilbourn (sometimes "Killbourn") was the son of Norman and Lucy (Peck) Kilbourn of Litchfield. Born on December 27, 1815, George Kilbourn was two years younger than Virgil Goodwin.[23] It is likely that all of the young men had been working together on the clocks for Henry Hart.

Meanwhile, the unfinished clock movements and parts that had been attached and placed in storage awaited resolution of the court case. The following information was found in the above-mentioned tally of the plaintiff's court costs (see Figure 87):

Virgil C. Goodwin Dr
vs.
Henry Hart
April 4ᵗʰ To moving Clock parts &c from J.
1835   Norths to Wm Lymans myself and two
          Hands with a Team one day ------}4.00
To paid William Lyman for Storing and taking
care of same two years and five months up to
this time ------}   10.00
                                           $14.00

Goshen August 30ᵗʰ 1837 Mainers Ives [?] Constable[24]

William Lyman of Goshen, who stored the clock parts for nearly two and one-half years, had an interesting connection with clock making. On March 19, 1834, he married Mary Ann Ives of Plymouth, the same woman who had unsuccessfully sued Silas B. Terry, the son of clockmaker Eli Terry, for breaking several promises to marry her a few years previously.[25]

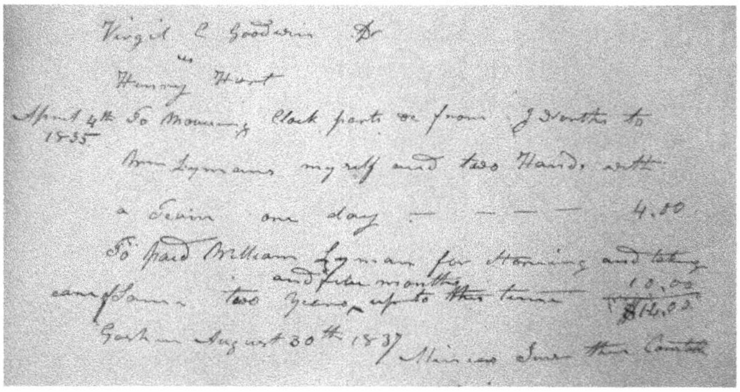

Figure 87. Account of expenses for moving and storing clock parts, included as part of the calculation of court costs in the case of Goodwin vs. Hart. *(Courtesy Connecticut State Library.)*

A reasonable speculation is that Dan H. Goodwin, Virgil C. Goodwin, and George Kilbourn had all become acquainted with one another, and with Edward Seymour, in one of the clock factories of Plymouth. Before the court case ended, the three young men left Goshen together, removing to Unionville, where it appears they all found employment with Seymour's clock-making firms. Dan Goodwin, said to have been an early employee of Seymour, Williams & Porter, reportedly later worked in the Unionville cabinet shop of Joshua Brewer and Lambert Hitchcock.[26] Although Dan Goodwin's name did not appear in Seymour, Williams & Porter's extant account books, on May 4, 1844, in his report to the Farmington District Probate Court, Pomroy Strong, the trustee of Edward Seymour's assigned estate, debited the estate to "Time spent in settling book acct Dan H. Goodwin $2.50,"[27] supporting an argument that Goodwin had indeed once worked for one or more of Seymour's firms.

Dan H. Goodwin did not own land in Unionville at the time of his assessment on personal property in Farmington consisting of only one wagon and one watch in 1843, the year in which he married Sarah Fuller, daughter of Joseph and Amy (Cornish) Fuller of West Avon.[28] At least one local history mentions Dan H. Goodwin and his family as having lived in one of the two yellow houses on Keyes Street that had been built for employees of the old woolen mill, later the eastern clock factory of Edward Seymour's firms. On January 13, 1840, Timothy Cowles sold the two boarding houses, together with a portion of the land formerly associated with Seymour's clock factory complex (but excluding the factory sites themselves), to

the clock-making firm of Williams, Orton, Prestons & Co.[29] Dan H. and Sarah Goodwin's daughter, Adelaide Emily Goodwin, is said to have been born in one of the boarding houses, on August 21, 1844.[30] This was a little over two years after the dissolution of the firm of Williams, Orton, Prestons & Co., and six years after the fire that had destroyed Edward Seymour's western clock factory building.

Wooden movement clock making had all but ended in Unionville village, when on June 18, 1846, Virgil C. Goodwin transferred one-quarter of an acre of land, with buildings standing thereon, to his brother Dan.[31] On the same day, Dan H. Goodwin mortgaged the parcel, bounded on the north by Virgil's then former clock-making partner, Russel Humphrey, east on a highway, south on Abraham Parsons, and west on Virgil Goodwin's own land, to E. O. Gridley to secure a $450 note to Gridley payable in two years, with annual interest, for "value received."[32] Dan Goodwin mortgaged the property a second time, also on June 18, 1846, to his brother Virgil, with the latter mortgage being "subsequent in its claims" to the mortgage held by Gridley.[33] At the time of the 1850 U.S. Census, Dan H. Goodwin was working at Unionville as a cabinetmaker. He was still living in Unionville in 1860, when his occupation was described as "mechanic." As of 1880, Dan H. Goodwin was employed in a local cutlery shop. He died on March 30, 1896.[34]

A half-brother of Dan H. and Virgil C. Goodwin, Phineas Bird Goodwin (1820–93), also settled in Unionville. "Bird" Goodwin appeared in the Town of Farmington's tax records in 1844, marrying Sarah Shattuck, daughter of Ansel and Maria (Burr) Shattuck in Unionville on October 10, 1847. Phineas B. Goodwin later reportedly served several terms as a local selectman and became involved in ice and coal dealing.[35] Whether he ever worked at clock making in the village is unknown.

In April 1835 shortly after moving to Unionville, Virgil C. Goodwin was admitted as an elector (voter) of Farmington. During the same year he began appearing on the town's tax lists, when he was assessed on personal property consisting solely of one watch. It is interesting that although the authors have seen no indication that he owned land in the village at the time, in 1837 the local selectmen approved Virgil Goodwin as a suitable person to keep a tavern.[36]

On July 12, 1839, Levi Smith, then "of Farmington," quitclaimed a half-acre parcel located in Unionville, described as a portion of "Lot No. 3 in the subdivision of Solomon Landgon's farm," to Virgil C. Goodwin. Smith's name was mentioned previously in Chapter 2 as having appeared in the account books of Seymour, Williams & Porter. Like Goodwin, he had probably been one of the firm's employees. How Goodwin obtained his interest in the parcel remains a mystery. The aforementioned quitclaim deed was not recorded until March 17, 1842,[37] nearly three years after it was signed.

In 1839, about the time that Levi Smith executed the quitclaim deed to Virgil C. Goodwin, shelf clocks with Goodwin's own labels were beginning to appear, and Smith was preparing to move to Bristol. There he began manufacturing 30-hr. wooden clock movements.[38] It was not previously known that Smith had worked at clock making in Unionville. Smith had purchased Lot No. 3, bounded south by the Farmington River, east and west by lands of Timothy Cowles, "formerly owned by Edward Seymour and others," and north by the "highway" leading to Farmington village, from grantors William Mather of Simsbury and Edward T. Hillyer of Granby, on February 6, 1838.[39] The portion of Lot No.

3 that Goodwin obtained from Smith was located on the parcel's southernmost edge, adjacent to the river, suggesting that its use was connected with transportation.

The emergence of small numbers of 30-hr. wooden movement shelf clocks bearing Virgil C. Goodwin's labels identifying Unionville as their place of origin, seems to have had something to do with the aftermath of the firm of Seymour, Hall & Co. In October 1837 the latter firm's machinery, tools, and finished and unfinished stock of clocks and parts had been placed under attachment in a lawsuit brought by C. D. Cowles. A few months later, on April 2, 1838, the firm's movement factory was destroyed by fire. It is speculated that Virgil C. Goodwin acquired some of the unfinished stock from Seymour's former (eastern) case shop, and commenced casing clocks, shortly after C. D. Cowles withdrew his suit against the partners in the former firm of Seymour, Hall & Co. in November 1838.

In fact, labels of the "early phase" of clocks produced by Goodwin suggest a connection with Edward Seymour's former firms. An example in a half-column and splat-style case, shown in Figures 88A–88C, bears

Figure 88A. Thirty-hr. wooden movement shelf clock by Virgil C. Goodwin, Unionville, exterior view.

Figure 88B. Label of clock shown in Figure 88A.

Figure 88C. H. Welton & Co. movement inside Virgil C. Goodwin clock shown in Figures 88A–88B. (See text.) (Figures 88A–88C courtesy of Chris Brown; Gene Gissin photos.)

a label that reads: "PATENT / CLOCKS / (*With Brass Bushings*,) / INVENTED BY ELI TERRY, / MANUFACTURED AND SOLD / BY / VIRGIL C. GOODWIN, / UNIONVILLE, CONN. / AT WHOLESALE AND RETAIL". In Chapter 2 it was noted that the phrase: "INVENTED BY ELI TERRY" was a consistent feature of the clock labels of the firms of Seymour, Williams & Porter and Seymour, Hall & Co. It is likely that this label wording reflected a licensing agreement dating to February 1832, between Seymour, Williams, & Porter and Eli Terry, in which the former firm paid Terry a fee of $100 to use the latter's wooden shelf clock movement patent. As mentioned in Chapter 2, the $100 payment to Terry for the use of his patent was recorded in Seymour, Williams & Porter's accounts.

As indicated above, clocks with Virgil C. Goodwin's own labels seem to date to two distinct periods, with the earlier group represented by the example shown in Figures 88A–88C. The label printer of this group, Hurlbut & Williams of Hartford, appeared in Geer's *Hartford City Directory* only for the year 1839, providing a reasonable estimated production date. Thirty-hr. wooden movements (Type 6.111) observed in these clocks are thought to have been made by H. Welton & Co.,[40] which firm had only recently commenced making wooden movements in the Terryville section of Virgil Goodwin's native town of Plymouth in 1839. An individual identified only as "Goodwin" appears in H. Welton & Co.'s extant accounts, now in the collection of the American Clock & Watch Museum.[41] Finding 30-hr. wooden movements only marginally profitable, H. Welton & Co. ceased producing them about 1842, but continued selling them from inventory for several years afterward. We will return to the matter of the second group of Goodwin's clocks later on.

In May 1841 the Farmington District Probate Court appointed Virgil C. Goodwin and Edward Seymour to appraise finished and unfinished clocks and other property in the assigned estate of the Unionville clock-making firm of Williams, Orton, Prestons & Co. Shortly thereafter, Goodwin began acquiring additional land in the village. On September 3, 1841, he purchased a parcel from William Griswold, bounded north on Griswold, south on Luther T. Parsons, west by Ephraim Fuller, Benjamin C. Hosford, and Frederick W. Crum, and east on a highway, "near the dwelling house of Asa B. Darrow,"[42] in close proximity to the former screw manufactory building, occupied since late 1835 by Williams, Orton, Prestons & Co. We have met Crum and Darrow

in previous chapters; both feature prominently in the next. By June 18, 1846, Virgil C. Goodwin's home "and other buildings" were situated on a nearby one-acre lot, bounded north by Joshua Brewer and Russel Humphrey, east by Dan H. Goodwin, south by Abraham [or Abram] Parsons, and west by a highway.[43]

On November 21, 1842, Goodwin mortgaged two parcels of land to George Richards, a creditor of Williams, Orton, Prestons & Co.'s assigned estate, to secure a note payable to Richards in 90 days. One of the witnesses to the deed was Cornelius R. Williams, a principal partner in Williams, Orton, Prestons & Co., which had dissolved in the spring of the same year.[44]

Figure 89. Exterior view of 30-hr. wooden movement shelf clock by Goodwin & Frisbie of Unionville, lower glass missing. An overpasted section of the label visible in the photo reads: "With Brass Bushings". (Courtesy Ken Snodderly; Ken Snodderly photo.)

It seems significant that the assignment of Williams, Orton, Prestons & Co.'s estate in May 1841, and the auction of its clock-making tools and machinery in January 1842, occurred about the same time that Virgil C. Goodwin became involved in two short-lived Unionville clock firms: Goodwin & Frisbie and Goodwin & Humphrey, probably in that order. What is known of the histories of these firms is recounted below.

Goodwin & Frisbie, Unionville

A very small number of stenciled half-column and splat-style, and a few beveled case clocks, have been seen, mostly incomplete, that bear the labels of the firm of Goodwin & Frisbie of Unionville. Intact examples seem to contain 30-hr. wooden movements of Type 6.111, mentioned above in connection with Virgil C. Goodwin's clocks, but produced by H. Welton & Co. of Plymouth.[45] Like Virgil C. Goodwin alone, Goodwin & Frisbie was probably involved in clock casing and sales, rather than movement making.

The exterior of a Goodwin & Frisbie shelf clock in a refinished half-column and splat-style case, with its bottom glass missing, is shown in Figure 89. A portion of the clock's label is overpasted with a small strip of paper, on which is printed the phrase: "With Brass Bushings". Figure 90 shows the label of a different example of a Goodwin & Frisbie clock, with no overpaste, and no mention of brass bushings. It is interesting that neither of the labels mentions Eli Terry. It is suggested that after Terry's patent protections expired in 1837, there was no longer any need to pay a licensing fee to Terry, or to credit him for his inventions.[46] Based on city directory data, the printer of the label shown in Figure 90 (barely visible), Joseph Hurlburt of Hartford, was active in 1838 and then again in 1840 through 1844, consistent with an estimated time frame of "about 1840" for Goodwin & Frisbie's clocks.

Isaac Porter Frisbie (1809–73), the second named partner in the firm of Goodwin & Frisbie, was born in Harwinton to parents Isaac Jr. and Cynthia (Rosseter) Frisbie and may have been a relative of Noah Preston and Eli Dewey Preston, two of the partners in Williams, Orton, Prestons & Co. Like Virgil C. Goodwin, Frisbie had likely been involved with clock making long before his arrival in Unionville: Frisbie's wife, the former Esther Roberts, was a daughter of clockmaker Wyllys Roberts (1798–1841) and a granddaughter of wooden movement tall clockmaker Gideon Roberts (1749–1813) of Bristol.[47]

On April 1, 1833, Isaac P. Frisbie was admitted as an elector in the Town of Farmington "By Certificate...

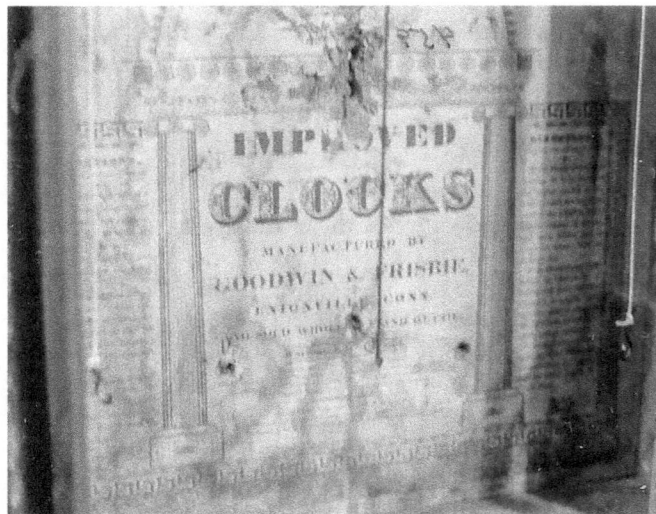

Figure 90. Label from a Goodwin & Frisbie, Unionville, 30-hr. wooden movement shelf clock. (Ward Francillon photo.)

from Windsor." It may be significant that several individuals soon to play prominent roles in the history of clock making in the villages of Farmington and Unionville were admitted as electors in the town of Farmington on the very same day: Cornelius R. Williams, Henry Tolles, and William L. Gilbert, all "by Certificate" from Bristol; George Marsh "by Certificate" from Litchfield; and Levi Smith "by Certificate" from Plymouth.[48] This information suggests that the men arrived as a group, having previously worked together in the Bristol clock factory of George Marsh and William L. Gilbert. If so, it would appear that they had also worked for George Marsh's firms at Cowles, Deming & Camp's mill, and perhaps for the Marsh firms' successors, Orton, Preston & Co., and then Williams, Orton, Prestons & Co., at Farmington village. The last mentioned firm moved to Unionville late in 1835. As explained in Chapter 1, Henry Tolles and Cornelius R. Williams played important roles in the Marsh firms and their successors.

Some four years before the clocks with Goodwin & Frisbie's labels appeared, on June 25, 1836, John T. Norton, Abner Bidwell, and James and Augustus Cowles, all of Farmington, deeded a parcel located in close proximity to the former Patent Wood Screw Manufactory (then occupied by Williams, Orton, Prestons & Co. as its clock factory), to Isaac P. Frisbie. The parcel, bounded north by land of the grantees, east by William Griswold, south by John Thompson, and west by a highway, was "100 links wide on the front and 31 links wide on the rear of said lot."[49] It is likely that the deed was part of Norton, Cowles, & Bidwell's plan to furnish or develop housing for workers in its nearby factory building. However, Frisbie did not remain in Unionville for very long afterward, disappearing from Farmington's tax lists beginning in 1841.[50]

Two additional bits of evidence connect Isaac P. Frisbie with the clock-making firm of Williams, Orton, Prestons & Co., perhaps as an employee. The first is that during December 1841, Frisbie's name appeared on the list of the approved creditors of Williams, Orton, Prestons & Co.'s assigned estate. Second, Frisbie's name appeared alongside those of Noah Preston, Henry C. Bull, and Heman H. Orton, three of the partners in Williams, Orton, Prestons & Co., on an extant list of Farmington subscribers to Connecticut's earliest anti-slavery newspaper, *The Charter Oak*, in 1838.[51]

It appears that after the end of Edward Seymour's Unionville firms in 1837 or 1838, and the dissolution of Williams, Orton, Prestons & Co. in early 1842, as many of Unionville's former clock factory employees struggled to find employment, a number of them formed small partnerships, purchasing or taking their severance pay in small quantities of leftover stock, and (even as clocks with wooden movements were rapidly becoming obsolete), working to dispose of completed clocks. Likely one such firm, Goodwin & Frisbie was short-lived, appearing on the Town of Farmington's tax lists only for the year 1840, when it was assessed taxes totaling $5.10 for properties consisting of only "1 waggon 1 horse." It is interesting that the tax records included "Isaac P. Frisby" under "V.C. Goodwin's List." The latter assessment was levied on one clock, one poll tax, and most significantly, on a half-acre parcel of land bounded south on John Thompson, west by highway, north by Norton, Cowles & Bidwell, and east by William Griswold, undoubtedly the parcel Isaac P. Frisbie had purchased from Norton, Cowles & Bidwell in June 1836. This record was followed by an individual assessment for Virgil C. Goodwin, for the same items, plus the parcel bounded

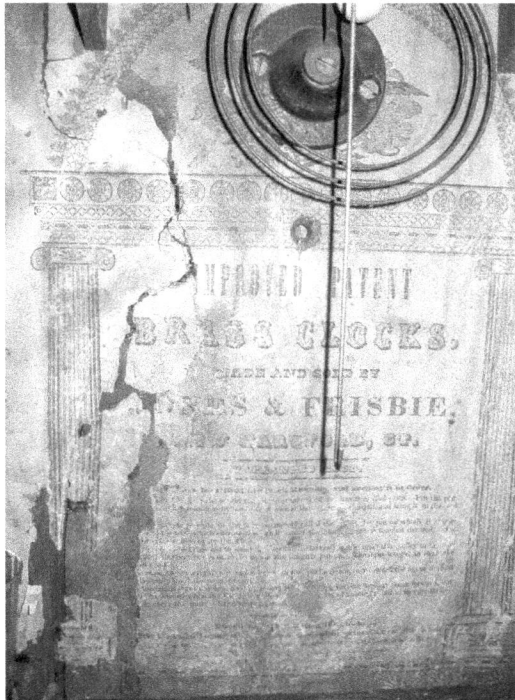

Figure 91. *Label from a 30-hr. weight-driven brass movement shelf clock by Jones & Frisbie, New Hartford, CT. (Courtesy Walt and Ethel Weber; MJD photo.)*

Figure 92. *Goodwin & Humphrey, Unionville clock in beveled case, door open, dial removed, showing label and 30-hr. wooden movement made by H. Welton & Co. Note brass bushings on movement. (Courtesy of Craig Sebold; Craig Sebold photo.)*

south on the Farmington River, in close proximity not only to the site of Edward Seymour's former clock-making facilities, but also to the sawmill and other buildings occupied as of 1840 by Williams, Orton, Prestons & Co.

On November 11, 1841, Isaac P. Frisbie, still of Farmington, deeded the half-acre parcel in Unionville he had purchased from Norton, Cowles & Bidwell in 1836, to Samuel E. Taylor (1813–?) of Farmington, who was also connected with Unionville clock making.[52] Dan H. Goodwin and Hiram Bascum [sic], both likely present or former clock factory employees of Unionville, witnessed the deed.[53] On February 18, 1841, Samuel E. Taylor, then of Colebrook, CT, married Juliet Orton (1821–?) of Farmington, a half-sister of Heman H. Orton, the "Orton" in Williams, Orton, Prestons & Co., at Farmington.[54] Later, in 1849, Samuel E. Taylor became a partner in the brass movement clock-making firm of Birge, Peck & Co. of Bristol. Taylor, who is said to have specialized in producing verges and other movement parts for clocks, continued to work at clock making in Bristol and Forestville, until 1873 or even later.[55]

No doubt surmising that the future of the clock industry lay in 30-hour movements stamped from sheet brass, sometime in 1841, Isaac P. Frisbie left Unionville to establish a clock-making partnership with one Henry Jones (1800–63), a prominent manufacturer, merchant, probate judge, property owner, and exporter of tin ware, at New Hartford, CT. The name of the new clock-making firm that produced 30-hr. weight-driven brass movement shelf clocks was Jones & Frisbie.

Henry Jones had once been a Farmington resident himself. Born at Barkhamsted (CT), he lost both parents at an early age. Subsequently, he was sent to Farmington to live with his guardian, Thomas Youngs. This was the same Thomas Youngs who, together with James and Augustus Cowles, John T. Norton, and Abner Bidwell, had been instrumental in developing the Patent Wood Screw Manufactory at Unionville.

Jones & Frisbie established their new clock factory in New Hartford's Pine Meadow district, on the south bank of Spruce Brook. The firm hired a number of workmen and peddlers and went on to produce clocks, which were marketed in the American South and West. However, losses in the South were said to have precipitated the firm's dissolution in 1844 or 1845.[56] Henry Jones seems to have been the firm's financier and sales manager; Frisbie may have been the principal clock maker or factory foreman. Ogee-style shelf clocks with 30-hr. brass movements bearing the labels of Jones &

Frisbie of New Hartford are known. The label of one such example is illustrated in Figure 91.

By 1850 Isaac P. Frisbie removed to Bristol, where he again found work as a clockmaker. At the time of the 1860 U.S. Census, Frisbie was maintaining residences in both Bristol and New York City, and still working at clock making. He died at New York City in 1873.[57]

Goodwin & Humphrey, Unionville

Shortly after the firm of Goodwin & Frisbie ended, Virgil C. Goodwin formed a second clock-making firm, once again apparently engaging in casing and sales, rather than movement manufacturing. Ogee and beveled case-style clocks bearing labels of the firm of Goodwin & Humphrey of Unionville are known. As in Goodwin & Frisbie's clocks, those of Goodwin & Humphrey contain 30-hr. wooden movements, apparently obtained from H. Welton & Co.[58] An unrestored example in a beveled case is shown in Figure 92. Its label reads: "IMPROVED / CLOCKS / With Brass Bushings / MANUFACTURED BY / GOODWIN & HUMPHREY, / UNIONVILLE, CONN. / AND SOLD WHOLESALE AND RETAIL." The clock's movement, also visible in Figure 92, is brass bushed. A second example, in an ogee-style case, is shown in Figures 93A and 93B. Consistent with the clock's label, its movement does not appear to be brass bushed.

The name of Virgil Goodwin's partner in the firm of Goodwin & Humphrey was associated with manufacturing in Unionville long after the wooden clock industry ceased to exist: Russel Humphrey. Humphrey was born at Farmington on December 7, 1812, a son of Captain Ralph Humphrey. His older brother, Montgomery, born in 1797, married Maria Gleason, the daughter of David Gleason of Farmington, in 1830. Montgomery Humphrey and copartner Asahel Gleason, doing business under the firm name of Humphrey & Gleason,[59] received local approval to keep "a house of public entertainment" in Farmington (Unionville) in July 1832. However, as local opinion shifted away from the use of "ardent spirits," the approval was suspended in January 1836.[60] At about the age of 21 in 1833–34, Russel Humphrey began appearing on Farmington's assessor's lists, when his local taxes were abated,[61] apparently for military service. He was admitted as an elector (voter) in 1835.[62] A founding member of the village's First Ecclesiastical Society in 1839,[63] Humphrey later became a deacon of Unionville's First Congregational Church,[64] which makes the following historical account all the more surprising.

On November 6, 1835, Farmington grand juror Erastus Porter appeared before Lemuel Whitman, Esq., a justice of the peace for Hartford County, to present a criminal complaint. The nature of the complaint was described in the court case file as follows:

> ...[O]n the fifth day of November 1835 at said Farmington, Russel Humphrey, Virgil C. Goodwin, and George Kilbourn, all of said Farmington, did, with force and arms, a violent and unprovoked assault make upon the person of John S. Hiscox of said Farmington, then being in the peace, and then and there conduct themselves in a violent and tumultuous manner with their hands and fists, and bruise and wound him the said Hiscox, and tear the clothes from his body and otherwise evilly entreat him the said Hiscox: all which doings of them...are against the peace and in violation of the statute of this State... Dated at Farmington this 6th day of November AD 1835.[65]

All of the men were undoubtedly acquainted with one another, and all were probably clock factory employees at the time. Virgil C. Goodwin and George Kilbourn had only recently arrived in Unionville from Goshen, where they had evidently worked together in the clock factory of Henry Hart. John S. Hiscox was assessed as a "mechanic" on the Town of Farmington's tax list for the year 1835.[66] In fact, Goodwin, Humphrey, and Hiscox had all enrolled as local electors on the same day during the previous spring.[67]

On the day of the grand juror's complaint, Justice of the Peace Whitman issued an arrest warrant (a portion of which is shown in Figure 94), reading (in part) as follows:

> ...forthwith to arrest the bodies of the said Russel Humphrey, Virgil C. Goodwin, and George Kilbourn, and them have, as soon as may be, before Horace Cowles, Esq. a Justice of the Peace for said County at his office in said Farmington, or before some other proper authority in said town, then and there to make answer to the foregoing complaint, and to be dealt with as to law and justice shall appertain. And you are also to summon David Cadwell, George Pierpont, Ephraim Porter, and John S. Hiscox, all of said Farmington, to appear before said court,

Figure 93A. Goodwin & Humphrey, Unionville, 30-hr. wooden movement shelf clock in ogee case, door open, showing label.

Figure 93B. Thirty-hr. wood, weight movement by H. Welton & Co., in clock by Goodwin & Humphrey shown in Figure 93A. (Figures 93A–93B courtesy Bob Higgins.)

at the time and place of trial, then and there to testify what they may severally know concerning the facts alleged in said complaint.[68]

On November 7 Pomroy Strong, a Farmington constable, arrested Russel Humphrey, Virgil C. Goodwin, and George Kilbourn, and also served the witnesses with their subpoenas, except for George Pierpont, who could not be found (see Figure 95). All of the men were scheduled to appear before the justice of the peace on November 14, 1835, at "9 o'clock in the forenoon."

When Humphrey, Goodwin, and Kilbourn were brought before Farmington Justice of the Peace Horace Cowles, each was charged with assault. Because George Kilbourn was a minor under 21 years of age, Asahel H. Lewis of Farmington was appointed to serve as his guardian during the trial. The three defendants "severally" pleaded not guilty. The justice of the peace, however, found them all guilty and fined them $7 each, to be paid to the Town of Farmington's treasury. The defendants were also ordered to pay the costs of the State's prosecution, $10.26, plus taxes thereon. The three young men (to whom $7 was undoubtedly a severe penalty), appealed to "the next County Court to be holden at Hartford...on the second tuesday of November AD 1835," when and where they again pleaded not guilty. Although they did not succeed in overturning the lower court's decision, the County Court reduced their fines to $1 each, which was then about one day's wages for an experienced factory worker.[69] Sometime after the incident occurred, John S. Hiscox left Farmington: the 1840 U.S. Census found him living in nearby Berlin, Connecticut.

On October 12, 1841, Russel Humphrey wed Aurelia Driggs of New Hartford, at Barkhamsted, CT. The firm of Goodwin & Humphrey was assessed taxes for property owned in Farmington only in 1841 and 1842, suggesting that perhaps the firm was established in the wake of the assignment of Williams, Orton, Prestons & Co.'s estate the previous spring,[70] to help complete and dispose of work in progress at that time. In 1841 the firm's property consisted of only one horse and one wagon. By 1842 Goodwin & Humphrey's taxable property consisted of two wagons. In addition, during both years the firm paid an "assessment" or business tax.[71] Thereafter, the firm disappeared from Farmington's tax lists, suggesting that it dissolved sometime in 1842. Consistent with a date of 1841–42 for the firm of Goodwin & Humphrey, the firm's clock labels were produced by printer Joseph Hurlburt of Hartford.[72]

Russel Humphrey was assessed taxes in Farmington for the year 1843 but again received an abatement. At that time, Humphrey owned no dwelling houses, but did own a three and one-half acre parcel of land in Unionville.[73] He had purchased the parcel, bounded north on William Griswold, east on a highway, south on property of Virgil Goodwin, and west on Virgil Goodwin, Frederick W. Crum, Ephraim Fuller, and Benjamin C. Hosford, and "near the dwelling house of Asa B. Darrow,"[74] from Virgil C. Goodwin on August 29, 1843, for $250. The parcel was situated near the former Patent Wood Screw Manufactory building, which continued to be designated as the "clock factory" on the Town of Farmington's tax lists until 1848. Therefore, it appears that Humphrey had been working in the

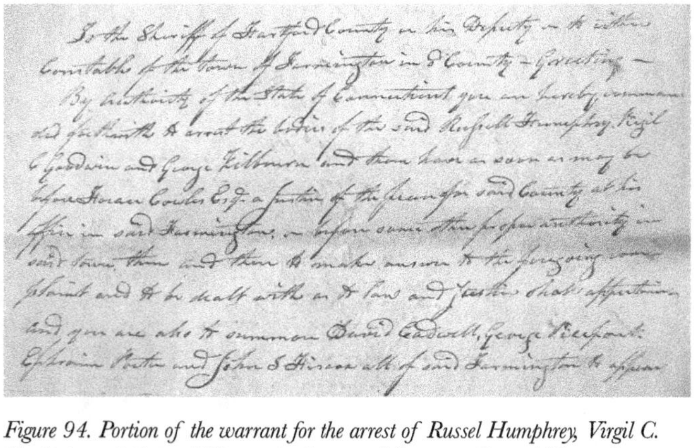

Figure 94. Portion of the warrant for the arrest of Russel Humphrey, Virgil C. Goodwin, and George Kilbourn, in the case of State vs. Russel Humphrey et al. (See text.) (Courtesy Connecticut State Library.)

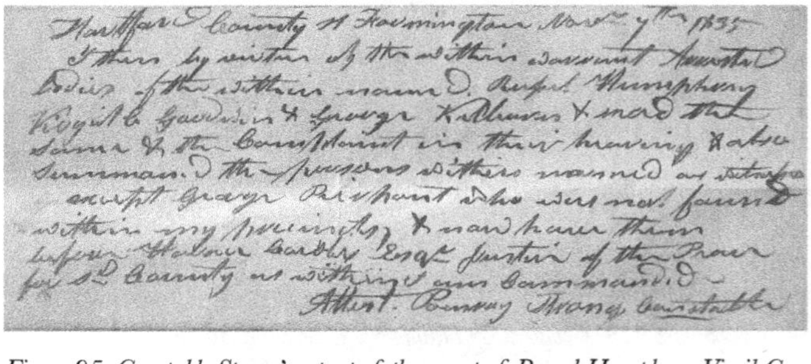

Figure 95. Constable Strong's report of the arrest of Russel Humphrey, Virgil C. Goodwin, and George Kilbourn, on November 7, 1835, in the case of State vs. Russel Humphrey et al. (Courtesy Connecticut State Library.)

building, and perhaps living with his family in nearby factory housing.

On November 6, 1842, the firm of Goodwin & Humphrey appeared in the extant account book, dating to the period 1842–96, of George Payne, a farmer and blacksmith of Unionville. This was the only entry for the firm, which purchased "loads wood delivered Notrop [sic]" and for which Payne received "goods to balance," specifically including "Mittens and shoes."[75] Perhaps it is not a coincidence that mittens and shoes were among the property of the Unionville firm of Williams, Orton, Prestons & Co. that had been sold, along with the firm's clock-making tools and machinery, at the auction held in January 1842.

After his involvement with clock making at Unionville ended, Russel Humphrey continued to work as a manufacturer in the village. On November 18, 1848, he deeded machinery "formerly owned by Allen Porter," including "...one pressing mill, one rolling mill, one plaining machine [sic]; one heavy stamp...together with my right of the shafts & belts connected with the machining for manufacturing iron spoons. Also one shop for tinning spoons, with the fixtures; said shop standing on the land of Norton, Cowles, & Bidwell,"[76] to Luther T. Parsons. The latter shop, described elsewhere as "small,"[77] seems to have been located adjacent to, if not within, the former Patent Wood Screw Manufactory building,[78] which had also been used from the latter part of 1835 until 1847 as a clock factory. How long Humphrey had been renting the above shop is unknown.

Russel Humphrey continued to manufacture metal spoons, washers, and rivets at Unionville as late as 1870, when his son took over the business.[79] The elder Humphrey died at Avon, CT, on December 28, 1873, at the age of 60 years.[80]

The Aftermath of Virgil C. Goodwin's Firms

Virgil C. Goodwin's involvement with clock making continued for some time beyond the probable end of the firm of Goodwin & Humphrey in 1842. As late as 1844–45, he was still dealing in clocks. On October 30, 1845, a deputy sheriff of Farmington summoned Goodwin to appear before the Hartford County County Court on the "Second Tuesday of November 1845[,] then and there to answer unto John Hunt of said Farmington."[81] Hunt, a prolific clock casemaker from Plainville village (also within the town of Farmington),

whose name appeared on the labels of shelf clocks containing both wood and brass movements produced during the 1830s and '40s, is known to have traded cases for movements produced by other manufacturers.

In his suit Hunt claimed that Virgil C. Goodwin owed him $70 on his book accounts. On the same day as he was served with the summons (i.e., October 30), a sheriff's deputy attached as Goodwin's property: "... one Tan Horses [sic] and double Wagon...."[82] Evidently, Goodwin was able to pay the debt: Hunt withdrew the suit during the March 1846 term of the Hartford County County Court.

More compelling evidence of Virgil C. Goodwin's involvement with clock assembly and sales (if not movement making) as late as 1844 or later, is found in the records of yet another lawsuit, this time brought by clockmakers Theodore Terry of Bristol, and Franklin C. Andrews of New York City, "partners under the name of Terry & Andrews."[83] According to the court case records, between March and June of 1843 and again in May 1844, Goodwin made several purchases of clock-related items from Terry & Andrews. These included "wire bells" in quantities of as many as 300 at one time, tablets, and a few "common brass clocks" (probably weight-driven, 30-hr. "ogees"). It is intriguing that Goodwin had also purchased quantities of as much as 30 lbs. of rolled brass and brass castings from Terry & Andrews at one time, for which he failed to remit payment, suggesting that he may have been assembling a few shelf clocks with brass movements. In any event, on November 30, 1847, the clerk of the Hartford County County Court commanded the County sheriff, "or either of the Constables of the Town of Farmington" "...that of the goods, chattels, or lands of the said debtor [i.e., Virgil C. Goodwin] within your precincts, you cause to be levied, and the same disposed of or...paid and satisfied unto the said creditors, the aforesaid sums [i.e., damages and costs of suit] being 84 dollars and 94 cents...with 17 cents more for this writ...."[84] This execution in favor of Terry & Andrews remained unsatisfied on February 8, 1848, when the Court issued a second execution against Goodwin in the matter.

A single beveled clock case bearing the label of "[V]. C. GOODWIN, / UNIONVILLE, CT." is known (see Figures 96A and 96B). Its weight-driven movement, now missing, was made of wood. Its label appears to be of a relatively late style. The label's printer, J. G. Wells of Hartford, was active in 1843, and then again from 1845 through 1847.[85] The existence of the clock case attests to Virgil C. Goodwin's involvement with clock making under his own name during a second (later) period, after the end of his respective partnerships with Isaac P. Frisbie and then Russel Humphrey.

Subsequently, Virgil C. Goodwin became involved in an additional partnership that may or may not have included clock sales. In a deed dated June 22, 1848, Goodwin transferred his interest in a piece of land in Unionville to Abner S. Whittlesey of Farmington for $100. The deed, which was subsequent in its claims on the land and buildings thereon to mortgages held by Luther T. Parsons, and to Luther T. Parsons and George Richards, included "all the interest I have in the livery stable and grocery business in Unionville under the name & firm of Whittlesey & Goodwin."[86] By 1850 Virgil C. Goodwin was living in New Britain (CT), where he had begun working as a locksmith.

The year 1851 found Virgil C. Goodwin a shareholder in the Wolcottville Hardware Manufacturing Co., which firm manufactured and sold goods made from iron, steel, wood, and brass, in the village of Wolcottville, within the town of Torrington, CT. This firm reportedly occupied a portion of "the old Alvord carriage shop" on the Naugatuck River's east branch.[87] One of the other stockholders in the firm was a man named George D. Wadhams (1800–62), a manufacturer of metal ware and buttons, who like Virgil C. Goodwin, had been (but perhaps still was) involved in clock making and sales.[88]

In 1857 the nation's economy once again fell under the shadow of a severe economic depression, which particularly impacted the manufacturing sector. At the time of the 1860 U.S. Census, Virgil C. Goodwin was living in Rochester, NY, where he was employed in selling sewing machines. Still living at Rochester in 1871, Goodwin served as a vestryman in a local church.[89] Sometime later, he traveled to Fox Lake, WI, "for renewal of health," but he died there on October 5, 1877, at the

Figure 96A. Empty beveled clock case with the label of "[V]. C. GOODWIN, UNIONVILLE, CT."

Figure 96B. Relatively late label of empty V. C. Goodwin clock case shown in Figure 96A, J. G. Wells, Hartford, printer. (See text.) (Figures 96A and 96B courtesy of The Unionville Museum.)

age of 64. His body was brought back to Unionville for burial.[90]

In our next and final chapter we will continue our discussion of "minor" clockmakers and clock-making firms of Unionville village, as a whole comprising a versatile and interconnected group. Appearing relatively late in the history of America's wooden movement shelf clock industry, they were forced to compete with increasingly large, well-capitalized firms. Furthermore, as in the case of Virgil C. Goodwin and his firms, it is probable that their brief existences stemmed from the breakup of Edward Seymour's firms and that of the firm of Williams, Orton, Prestons & Co.

Notes and References

1. Kenneth D. Roberts and Snowden Taylor. *Eli Terry and the Connecticut shelf clock*. 2nd ed. Fitzwilliam, NH: Ken Roberts Publishing Co., 1994.

2. Nelson Burr. *The early labor movement in Connecticut 1790–1860*. West Hartford, CT, 1972.

3. James J. Goodwin. *The Goodwins of Hartford, Connecticut*. Hartford, CT: Brown and Gross, 1891.

4. *Goodwin vs. Hart*. Litchfield County Superior Court Files, Box 218, RG3, Connecticut State Library.

5. Alfred Andrews. *Genealogical history of Deacon Stephen Hart and his descendants 1632–1875*. Hartford, CT: Alfred Hart, 1875.

6. A. G. Hibberd. *The history of the town of Goshen, Connecticut*. Hartford, CT: Case, Lockwood, Brainard & Co., 1897: 369.

7. Samuel Orcutt. *History of Torrington, Connecticut*. Albany, NY: J. Munsell, printer, 1878.

8. Research files of the American Clock & Watch Museum, Bristol, CT.

9. Alfred Andrews, *Genealogical history of Deacon Stephen Hart*, 1875.

10. Samuel Orcutt, *History of Torrington, Connecticut*, 1878.

11. An additional Goshen firm that produced wooden movement shelf clocks ca. 1831–36 was Northrup & Smith. See: Sonya L. Spittler, Thomas J. Spittler, and Chris H. Bailey. *Clockmakers and watchmakers of America by name and by place*. 2nd ed. Columbia, PA: National Association of Watch & Clock Collectors, 2011.

12. Snowden Taylor. "The account book of Alpha Hart, Goshen, CT 1815–1838." *Timepiece Journal of the American Clock & Watch Museum*, Vol. 5, No. 3 (Spring 1993): 62.

13. *Goodwin vs. Hart*.

14. Ibid.

15. Ibid.

16. Ibid.

17. Ibid.

18. Ibid.

19. "Oyer" is defined as "a copy of a promissory note or other written obligation sued upon." (See: James D. Folts. *Duely & Constantly Kept*. Albany, NY: New York State Court of Appeals and The New York State Archives and Records Administration, 1994:74).

20. *Goodwin vs. Hart*.

21. Ibid.

22. Plymouth Vital Records, Vol. 1, p. 11. See also: James J. Goodwin. *The Goodwins of Hartford, Connecticut* 1891.

23. Payne K. Kilbourn. *The history and antiquities of the name and family of Kilbourn*. New Haven, CT: Durrie & Peck, 1856.

24. *Goodwin vs. Hart*.

25. M.J. Dapkus. "So Silas B. Terry Had a *Past...*" NAWCC *Bulletin*, No. 401 (January/February 2013):89–92.

26. Mabel Hurlburt. *Farmington town clerks and their times*. Hartford, CT: Finlay Press, 1943.

27. Farmington Probate District Estate Papers, Connecticut State Library.

28. James J. Goodwin, *The Goodwins of Hartford, Connecticut*, 1891.

29. Farmington Land Records (FLR) Volume (V.) 47, pp. 290–291, January 13, 1840.

30. Mabel Hurlburt, *Farmington town clerks and their times*, 1943.

31. FLR V. 49, p. 153, June 18, 1846.

32. FLR V. 49, p. 154, June 18, 1846.

33. FLR V. 49, p. 156, June 18, 1846.

34. LDS Family Search website (online source).

35. James J. Goodwin. *The Goodwins of Hartford, Connecticut*, 1891; and 1870 U.S. Census.

36. Mabel Hurlburt, *Farmington town clerks and their times*, 1943.

37. FLR V. 45, p. 397, July 12, 1839; recorded March 17, 1842.

38. Roberts and Taylor, *Eli Terry and the Connecticut shelf clock*, 1994.

39. FLR V. 46, p. 529, February 6, 1838. Smith sold the parcel, minus the portion quitclaimed to Goodwin, to Gideon Walker on April 5, 1839. (See: FLR V. 48, p. 8).

40. Roberts and Taylor, *Eli Terry and the Connecticut shelf clock*, 1994. See also: Snowden Taylor. "More on H. Welton & Co. wood movements." NAWCC *Bulletin*, No. 227 (December 1983): 737–738. In an unpublished update to the latter, author ST proposes renumbering these movements as Type 8.121.

41. Snowden Taylor. "H. Welton & Co., 1839–1846, Part 4." *Timepiece Journal of the American Clock & Watch Museum*, Vol. 2, No. 8 (Winter 1983):171–192. Neither Goodwin's first name nor the nature of his transactions with H. Welton & Co. was specified in the firm's accounts.

42. FLR V. 48, p. 280, September 3, 1841.

43. See, for example: FLR. V. 49, p. 155, June 18, 1846; and FLR V. 49, p. 259, May 3, 1847.

44. See Chapter 4. Furthermore, according to the *Connecticut Courant* of June 18, 1842, Cornelius R. Williams and Virgil C. Goodwin had each recently been appointed to one-year terms as justices of the peace for the town of Farmington.

45. Roberts and Taylor, *Eli Terry and the Connecticut shelf clock*, 1994: 347–352.

46. M.J. Dapkus. "A new look at Eli Terry's patent infringement lawsuits." NAWCC *Bulletin*, No. 401 (January/February 2013): 60–68. Because the applicable Terry patents were actually replacements issued in 1826, it is possible that some confusion existed within the clock-making community about whether they instead expired in 1840.

47. Sonya L. Spittler, Thomas J. Spittler, and Chris H. Bailey. *Clockmakers and watchmakers of America by name and by place*, 2011. The maiden names of the mothers of Isaac P. Frisbie and the Prestons were the same.

48. Farmington Town Records, Vol. 2, Part 2, Farmington Tax Assessor's Office.

49. FLR V. 46, p. 379, June 25, 1836.

50. Town of Farmington Tax Records, RG62, Connecticut State Library; and 1840 U.S. Census, Farmington, CT.

51. "Charter Oak Prospectus," June 14, 1838, Farmington Public Library.

52. As of 1850 Samuel E. Taylor was living in Farmington (probably in Plainville village), when and where the U.S. Census recorded his occupation as "clockmaker."

53. FLR V. 45, p. 391, November 11, 1841.

54. Vital Records Index, Connecticut State Library; and 1850 U.S. Census, Farmington, CT.

55. Kenneth D. Roberts and Snowden Taylor. *Forestville clockmakers*. Fitzwilliam, NH: Ken Roberts Publishing Co., 1992: 169.

56. Sarah L. Jones. "New Hartford." In: *History of Litchfield County, Connecticut*. Philadelphia, PA: J. W. Lewis & Co., 1881.

57. Sonya L. Spittler, Thomas J. Spittler, and Chris H. Bailey. *Clockmakers and watchmakers of America by name and by place*. 2011.

58. Roberts and Taylor, *Eli Terry and the Connecticut shelf clock*, 1994: 347–352.

59. Frederick R. Humphrey. *The Humphreys family in America*. New York, NY: Humphreys Print, 1883.

60. Mabel Hurlburt, *Farmington town clerks and their times*, 1943.

61. Farmington Town Records (Selectmen's Records), Farmington Public Library.

62. Farmington Town Records, Vol. 2, Part 2, Farmington Tax Assessor's Office.

63. First Congregational Church Unionville, Records, Connecticut State Library.

64. Frederick R. Humphrey, *The Humphreys family in America*, 1883.

65. *State vs. Russel Humphrey et al.* Hartford County County Court Files, Box 427, RG3, Connecticut State Library.

66. Town of Farmington, Assessments 1835, Box 11, RG62, Connecticut State Library.

67. Farmington Town Records, Volume 2, Part 2, Farmington Tax Assessor's Office.

68. *State vs. Russel Humphrey et al.*

69. It appears unlikely that the assault was connected with the riot in Farmington sparked by an antislavery speaker in December 1835. (See: Christopher Bickford. *Farmington in Connecticut.* 2nd ed. Farmington, CT: Farmington Historical Society, 2008).

70. See Chapter 4.

71. Town of Farmington, Individual Tax Returns, Box 12, RG62, Connecticut State Library.

72. D. R. Slaght. "Printers of Hartford 1825 Thru 1860." Unpublished manuscript, NAWCC Library.

73. Town of Farmington, Individual Tax Returns, Box 12.

74. FLR V. 48, p. 434, August 29, 1843.

75. "Accounts of George Payne 1842–1896." Farmington Public Library.

76. FLR V. 49, p. 455, November 18, 1848.

77. FLR V. 53, p. 23, April 13, 1853. This deed was witnessed by James L. Cowles, author of "Unionville." In: J. Hammond Trumbull, ed. *The memorial history of Hartford County, Connecticut.* Vol. 2 of 2. Boston, MA: Edward Osgood, 1886, in which it was stated that Russel Humphrey made spoons in the former Patent Wood Screw Manufactory.

78. Betty Coykendahl. "History of Unionville Houses." Unpublished manuscript, 1979, Farmington Public Library.

79. Christopher Bickford. *Farmington in Connecticut.* 2nd ed., 2008.

80. Death Index, Connecticut State Library; and Farmington Vital Records V. 2, pp. 476–477.

81. *John Hunt vs. Virgil C. Goodwin.* Hartford County County Court Files, Box 449, RG3, Connecticut State Library.

82. Ibid.

83. *Terry & Andrews vs. Virgil C. Goodwin.* Hartford County County Court Files, Box 452; and Hartford County County Court Executions, Box 553, RG3, both in the Connecticut State Library.

84. *Terry & Andrews vs. Virgil C. Goodwin.* Hartford County County Court Executions, Box 553.

85. D. R. Slaght, "Printers of Hartford 1825 Thru 1860."

86. FLR V. 49, p. 437, June 22, 1848.

87. Samuel Orcutt. *History of Torrington, Connecticut,* 1878, 105.

88. Roberts and Taylor, *Eli Terry and the Connecticut shelf clock,* 1994, 359.

89. See: Farmington District Porbate Estate Papers for Virgil C. Goodwin; and F. De W. Ward. *Churches of Rochester: an ecclesiastical history of Rochester, New York.* Rochester, NY: E. Darrow, 1871.

90. James J. Goodwin, *The Goodwins of Hartford, Connecticut,* 1891.

Chapter 7

More "Minor" Clockmakers and Firms of Unionville Village: Crum & Barber; George Barber; Kilbourn & Darrow; Samuel E. Curtiss; and Levi Smith of Farmington and Bristol

Here we continue our discussion of "minor" clockmakers and firms of Unionville village. All used 30-hr. wooden movements (perhaps exclusively), and all are unusual because they appeared relatively late in the history of wooden movement clock making.

As we have seen, during the early 1830s, the remarkable success of Connecticut's clock industry had rested largely on shelf clocks with 30-hr. wooden movements.

Some clocks with 8-day wooden movements and some with 8-day cast or strap brass movements were also produced, but in far fewer numbers. Soon, however, abundant competition, combined with a persistent economic depression following the financial Panic of 1837, brought new challenges. Many clockmakers and firms, particularly those who had taken on heavy debt burdens, failed. By 1839 some of Connecticut's larger clock-making

Figure 97A. Beveled clock case by Crum & Barber, Unionville. The case once contained a 30-hr. wood movement, now missing. The clock's tablet may be a replacement.

Figure 97B. Label in Crum & Barber clock case shown in Figure 97A. (Figures 97A and 97B courtesy of The Unionville Museum, MJD photos.)

firms were no longer finding profit in 30-hr. wooden movements and began phasing them out.[1]

Sensing an opportunity linked with Connecticut's up-and-coming rolled brass industry, on June 27, 1839, clockmaker Noble Jerome (1801–61) of Bristol, the brother of Chauncey Jerome, received a U.S. Patent for a 30-hr. clock movement stamped from sheet brass. Although initially somewhat more expensive than its wooden counterpart,[2] the new brass movement had fewer limitations. Its introduction proved successful. Thereafter, many of Connecticut's remaining wooden clockmakers were forced to diversify, consolidate, or find new lines of work.

In light of these developments, the emergence of several small-scale clockmakers and firms in Unionville between 1839 and 1841—or in some cases even later—seems surprising. Nonetheless, markets for inexpensive wooden movement shelf clocks, regardless of their approaching obsolescence, persisted for a time, particularly in remote, rural areas. Moreover, as we will see, at least some of the "late" Unionville clock-making firms were established in response to specific circumstances, in the wake of the breakup of Edward Seymour's firms in 1838 and that of Williams, Orton, Prestons & Co. in 1842.

Crum & Barber, Unionville

For reasons that will be made clearer shortly, shelf clocks with Crum & Barber's labels were previously thought to date to 1835–44.[3] However, on the basis of the style(s) of their cases, and on their label and printer data,[4] it is more likely that these clocks date to 1841–44. Figures 97A and 97B show an incomplete example in a beveled case, a style that made its appearance about 1839. The clock's dial and 30-hr. wooden movement are missing. In fact, a search of Farmington tax assessor's records located the firm of Crum & Barber *only* on the town's 1842 list,[5] supporting an argument that it was established in 1841 or 1842, about the time the clock-making firm of Williams, Orton Prestons & Co. assigned its estate for the benefit of its creditors.

The first named partner in the firm of Crum & Barber was Frederick W. Crum. There can be little doubt that George Barber (sometimes spelled "Barbour") was his partner. We have met both men before; for example, in Chapter 2, they appeared in Seymour, Williams & Porter's accounts beginning in 1834 and 1835, respectively. Later, we encountered Frederick W. Crum as a partner in Williams, Orton, Prestons & Co. from its inception on or about

December 15, 1835, but leaving the firm sometime between September 1836 and August 1840. Finally, evidence presented in Chapter 4 documented Crum's subsequent involvement in Williams, Orton, Prestons & Co., once again as a partner, about the time the firm was broken up and portions sold off in 1842.

It is probable that Frederick W. Crum was already acquainted with Eli D. and Noah Preston (two of his future partners in Williams, Orton, Prestons & Co.), before his arrival in Farmington circa 1833. He was born in Winchester (CT) on March 21, 1813, the son of William and Hannah Crum.[6] His first wife, Ellice (sometimes "Ellis"], was born on March 4, 1812, to parents Levi and Ann (Langdon) Catlin, in the Prestons' hometown of Harwinton.[7]

Frederick W. Crum was still a partner in Williams, Orton, Prestons & Co., when on May 2, 1836, his wife purchased a small parcel of land in Unionville, "two roods and thirty seven rods" in aerial extent, from John T. Norton, Abner Bidwell, and James and Augustus ("J. & A.") Cowles,[8] the partners in Norton, Cowles, & Bidwell. These men owned the former Patent Wood Screw Manufactory, then occupied by Williams, Orton, Prestons & Co. as its clock manufactory. The parcel in question was located nearby. Such a sale was unusual: according to author Norris G. Osborn, in Volume 2 of his *History of Connecticut* (Albany, NY, 1925), prior to 1877, married women could not legally own property in Connecticut under their own names, and it is unclear why the parcel was deeded to Mrs. Crum instead of to her husband. Furthermore, the deed to Ellice Crum was not recorded until December 7, 1836,[9] the day on which the Crums mortgaged the same parcel, then described as having a dwelling house thereon, to J. & A. Cowles.[10] On May 4, 1840, Frederick W. Crum purchased an adjacent parcel in his own name, "bounded north on John Thompson, east on William Griswold, south on Ellice Crum, and west on highway, 56 links on the front or west and one chain 65 links on the east end or rear," also from Norton, Cowles & Bidwell,[11] thereby enlarging the Crums' home lot in Unionville.

As mentioned in Chapter 4, on July 3, 1890, about five years before his death, Frederick W. Crum's recollections of Unionville's clock industry were published in the *Bristol* (CT] *Herald*. The relevant portion of the article is presented again, below:

> ...F. W. Crum and Mr. Barbour formed a partnership about 1835. They bought the [Patent

Wood] Screw company's property, also that of Marsh & Gilbert, clock makers in Farmington and moved the business from Farmington to the Screw Shop. A company by the firm name of Williams, Orton & Preston was making eight-day weight brass clocks in 1836 and Crum & Barbour bought a portion of the business in 1841, Mr. [Heman] Orton purchasing the case department of that company and made cases for the movements manufactured by Crum and Barbour... The firm of Seymour, Williams & Porter was burned out in 1837 [sic; the actual date of the fire was April 2, 1838], but the firm of Crum & Barbour continued to make clocks here [i.e., Unionville] until 1844, when the price of clocks, which had formerly been $18 got so low that they went out of the business... Barbour went West and succeeded financially in the furniture business. Mr. Crum went on...manufacturing rivets... He afterward engaged in the manufacture of saws and for the last twenty-one years has been in the undertaking business...[12]

Without a doubt, Frederick W. Crum was both an active participant in, and an important eyewitness to, the history of clock making in Unionville. However, his brief recollections as summarized in the *Bristol Herald* leave much unexplained. For example, the article does not mention either that Crum and Barber probably supplied parts or labor to the clock-making firm of Seymour, Williams, & Porter beginning in 1834 (as the evidence in the firm's accounts suggests), or that Crum had been a founding partner in Williams, Orton, Prestons & Co. Furthermore, Crum's published recollections do not mention that, although the latter firm did produce a few 8-day brass movements (e.g., only about 300 in 1839),[13] Williams, Orton, Prestons & Co. was chiefly a high-volume producer of 30-hour and some 8-day wooden movement shelf clocks.

Frederick W. Crum's partner, George Barber, arrived in Farmington in 1834, at about the age of 20.[14] As related in Chapter 2, the clock-making firm of Seymour, Williams & Porter gave George Barber and Frederick W. Crum (individually) promissory notes in return for unspecified "value received," on March 23, 1835, and June 1, 1835, respectively.[15] Consequently, although the authors have seen no evidence that Crum and Barber were copartners at the time, it is reasonable to infer that each of the two men had been either employees of, or

suppliers to, Seymour, Williams & Porter.

Soon after the dissolution of the firm of Seymour, Williams & Porter in February 1835, it appears Frederick W. Crum and George Barber went their separate ways. A deed dated September 10, 1836, provides strong evidence that George Barber became employed by Seymour, Hall & Co., the successor firm to Seymour, Williams & Porter. In the deed, Edward Seymour (the principal partner in both firms), and Austin F. Williams (the firms' treasurer), transferred a parcel of land to Barber for only $1.00.[16] The parcel was one of five that Seymour and Williams had purchased at auction from Solomon Langdon's estate, for the lump sum of $1,400. It was bounded on three sides by Seymour and Williams' jointly owned properties.

As described in Chapter 3, Seymour, Hall & Co.'s clock-making activities were interrupted on October 7, 1837, when the firm's machinery and stock of unfinished clocks, cases, and parts were placed under attachment in a lawsuit brought by Chauncey D. Cowles over an unpaid debt. Within days, Seymour and Williams's clock factory lands and buildings were placed under attachment in several additional lawsuits over unpaid debts. On April 2, 1838, the firm's western (movement) factory was consumed by fire. Timothy Cowles, who held mortgages on the former Seymour, Williams, & Porter properties, acquired them for unpaid debts in June 1838. In November of the same year, C. D. Cowles withdrew his suit against Seymour, Hall & Co.; presumably, the firm regained its machinery and unfinished stock at that time. Afterward, the firm's remaining stock and equipment were likely sold to help pay its debts. In light of the other still pending lawsuits (particularly the suit brought by Joseph R. Gillet against Seymour, Hall & Co.), perhaps this needed to be done discretely.

On February 15, 1839, George Barber deeded the parcel he had obtained from Seymour and Williams, to William Bradley,[17] Seymour, Hall & Co.'s "successor." By July 1839, with Edward Seymour and his former partners in Seymour, Hall & Co. out of state (presumably collecting on notes taken in payment for clocks in Alabama), Barber was producing wooden movement shelf clocks at Unionville,[18] under his own labels.

Meanwhile, in December 1835, Frederick W. Crum joined the firm of Williams, Orton, Prestons & Co. Although the authors have seen no direct proof that either Crum or the firm of Crum & Barber, purchased equipment from Marsh, Gilbert & Co. or from the Patent Wood Screw Manufactory (as stated in Crum's recollections), neither scenario is improbable. For example, recalling that the firms of Orton, Preston & Co. and Williams, Orton, Prestons & Co. (in that order) were the successors to George Marsh's former clock-making firms of Farmington village, it would not be surprising if Crum, either acting alone or as a partner in Williams, Orton, Prestons & Co., obtained some of the former Marsh firms' clock-making equipment.

Additional circumstantial evidence consistent with Frederick W. Crum's recollections is found in a series of advertisements appearing in the *Connecticut Courant* beginning on February 4, 1837. The advertisements offered for sale equipment "lately owned by the Patent Wood Screw Manufacturing Company...with a variety of tools suitable for a machinist...,"[19] plus the factory clock, at Unionville (Figure 98). However, the equipment did not sell quickly in the precarious economic environment. An auction of the firm's remaining stock of unsold machinery, tools, and "implements for mechanics and machineists (sic], Iron, Steel, &c...,"[20] took place on August 29, 1837, and it is not inconceivable that, in accordance with Crum's published recollections, either he or Williams, Orton, Prestons & Co. would have purchased some items, either at the time the advertisements appeared or at the auction sale.

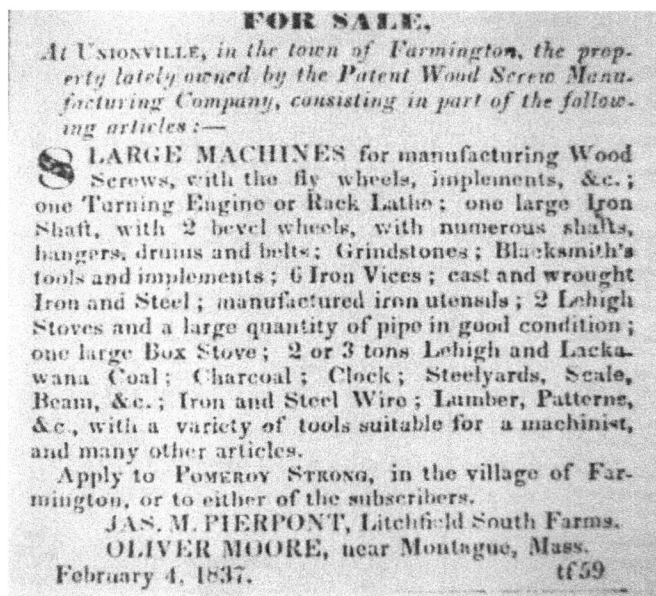

*Figure 98. Advertisement of property for sale belonging to the former Patent Wood Screw Manufacturing Co. of Unionville. (*Connecticut Courant, *February 4, 1837.)*

By October 30, 1839, George Barber may have been working for Williams, Orton, Prestons & Co., although it is uncertain whether Frederick W. Crum was still a partner in the latter firm at that time. On that day, a Hartford, CT, forwarding merchant gave Williams, Orton, Prestons & Co. its receipt, stating: "Received in store, Hartford...from Geo. Barber Seventeen cases clocks & five boxes Weights Marked E.N. Marvin to forward to New-York by the Steam Boat CLEOPATRA, (danger of seas excepted,) Unto S Holmes NY (signed] Wm Savage Agent."[21] As explained in Chapters 4 and 5, Ezra N. Marvin of Rockville, IN, was a longtime clock customer of Williams, Orton, Prestons & Co.

George Barber may have accompanied the clocks on their westward journey. He seems to have been absent from Unionville for the next one to two years,[22] returning to join Frederick W. Crum in forming the firm of Crum & Barber in 1841 or 1842. On July 1, 1843, Barber purchased Noah Preston's dwelling house in the village. This was the same day Noah Preston sold out his interest in the former firm of Williams, Orton, Prestons & Co. to his brother, Eli D. Preston,[23] supporting an argument that Crum & Barber did in fact succeed William, Orton, Prestons & Co. by purchasing a portion of its business. Barber retained the house until August 2, 1848, the year in which Norton, Cowles, & Bidwell's factory lost its designation as "the clock factory" on the local tax rolls. On that day, he sold it to David A. Keyes,[24] who was then manufacturing cutlery in Edward Seymour's former eastern clock factory building.

A few additional bits of circumstantial evidence suggest that the firm of Crum & Barber in some sense emerged as Williams, Orton, Prestons & Co.'s successor. For example, during 1841–42 Frederick W. Crum was identified as a creditor of Williams, Orton, Prestons & Co.'s assigned estate, although George Barber was not.[25] In fact, Crum and Chauncey Rowe were initially designated as the estate's trustees, but the two men resigned their trusteeships on May 6, and the Farmington District Probate Court appointed new individuals to take their places.[26] A few days later, on May 16, 1841, Frederick W. Crum assumed the role of Williams, Orton, Prestons & Co.'s "agent," likely in helping to finish and dispose of the unfinished clocks and movements in the latter firm's inventory, under the direction of Cornelius R. Williams.[27] Therefore, it appears that the firm of Crum & Barber was established either to assist in carrying out that purpose or as an outgrowth thereof.

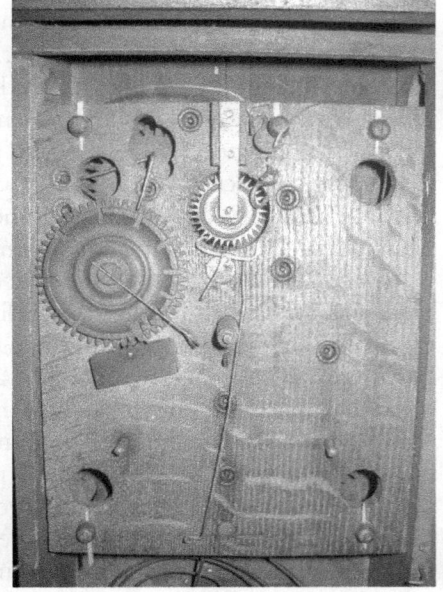

Figure 99A. Crum & Barber, Unionville, 30-hr. wood movement shelf clock in beveled case, exterior view.

Figure 99B. Label of Crum & Barber, Unionville, shelf clock shown in Figure 99A.

Figure 99C. Thirty-hr. wooden movement attributed to Levi Smith or L. Smith & A. Blakeslee, in Crum & Barber shelf clock of Figures 99A and 99B. (Figures 99A - 99C courtesy private collection, MJD photos.)

As mentioned above, a few shelf clocks bearing the labels of Crum & Barber, Unionville, are known. Most are housed in beveled cases, although (as we will see) the firm may also have produced clocks with ogee-style cases. A relatively intact bevel cased example appears in Figures 99A–99C. The clock's label (Figure 99B) refers to "EXTRA" clocks. Brass bushings are not mentioned, and the clock's movement (shown in Figure 99C) does not appear to be brass bushed. The movement's round pendulum suspension stud, visible in Figure 99C, is a feature associated with 30-hr. wooden movements made by Levi Smith or Smith's Bristol firms, and is not known to have been a feature of Williams, Orton, Prestons & Co.'s movements. Indeed, all known examples of Crum & Barber clocks, including the present example, contain movements attributed to Levi Smith, who produced relatively large numbers of 30-hr. wooden movements at Bristol, first in partnership with others and then alone, between 1840 and 1847.[28] This is in apparent conflict with Frederick W. Crum's published recollections, in which he seems to state that the firm of Crum & Barber produced its own clock movements. Nonetheless, if Crum & Barber did not purchase Williams, Orton, Prestons & Co.'s movement-making machinery at the

auction sale of January 1842, it would not be surprising that the firm turned to Levi Smith as a movement supplier.

Levi Smith arrived in Farmington sometime before April 1, 1833, when along with George Marsh, William L. Gilbert, Cornelius R. Williams, and others associated with the clock-making firms of George Marsh and their successors, he was admitted as an elector in the Town of Farmington.[29] Subsequently, Smith's name, along with the names of Frederick W. Crum and George Barber, appeared in the Seymour, Williams, & Porter accounts. Smith had been the purchaser of at least three parcels of land that were once part of Solomon Langdon's farm, all of which abutted lands of Edward Seymour and Austin F. Williams, and had served as a justice of the peace for the Town of Farmington during the 1830s. As mentioned above, Smith moved to Bristol in 1839 to start his own wooden movement-making and livery businesses.

As mentioned above, along with that of Frederick W. Crum, in December 1841, Levi Smith's name appeared on the list of the approved creditors of Williams, Orton, Prestons & Co. assigned estate. The nature of the firm's indebtedness to Smith (a resident of Bristol at the time), is unknown, although it could have been for 30-hr. wooden movements. More on Smith later.

In light of the information presented above, if Crum & Barber did in fact produce the movements for its own clocks (as suggested in Crum's published recollections), it appears that one or both of two possible scenarios must be true: (1) some of the movements hitherto attributed to Levi Smith were instead made by Crum & Barber, and (2) a reliable means of distinguishing the movements of Crum & Barber from those made by Levi Smith has yet to be discovered.[30]

Frederick W. Crum's published recollections mention an ending date of "about 1844" for the firm of Crum & Barber. Information supporting an 1844 ending date for the firm was found in the extant accounts, dating from 1842 to 1896, of a Unionville farmer and blacksmith named George Payne. Beginning in late 1842, Crum & Barber purchased items, including cord and slab wood, rafters, and "wood delivered to Goodwin's kiln dry" (likely referring to Virgil C. Goodwin), from Payne. Crum & Barber paid Payne in goods, including, for example, two pounds of coffee, a pencil case, and an "ogee clock" on October 12, 1842,[31] not only suggesting that Crum & Barber was in business as of that date but also suggesting that the firm may have produced a few shelf clocks with ogee cases. The last entry for Crum & Barber in Payne's accounts occurred on June 5, 1844, when the firm was credited by an order on Eli D. Preston. This information is consistent with Crum's recollection of the firm's 1844 ending date.

After his partnership with Barber ended, Frederick W. Crum remained in Unionville. Thinking about his future after clock making, on February 27, 1843, he and Eber N. Gibbs[32] purchased "one Hook machine, one eye machine, one pair of rolls for flatting, one vendse, one lathe, 700 lbs. of brass wire, 150 lbs. unfinished hooks eyes, 500 dozen boxes, and 2 boxes of clocks" from Charles H. Whittlesey and Orson L. Adams, both "late of Farmington, late partners under the name of Whittlesey & Adams..."[33] Next, on April 29, 1844, Crum purchased a 23-acre parcel in Unionville from George W. Payne, bounded south on Cornelius R. Williams, Eli D. Preston, and Albert Hill.[34] In June of the same year, Crum purchased a small parcel from William Griswold. This parcel, running from Lovely St. west to Crum's own land, was bounded south on Russel Humphrey, until recently Virgil C. Goodwin's partner in a clock casing and sales firm.[35] James Cowles quitclaimed a small parcel of land with a barn on it to Ellice Crum for $1

on January 31, 1845, the same day Frederick W. Crum deeded the parcel to Cowles for $30.[36] On August 16, 1845, Crum mortgaged the parcel he purchased from George W. Payne to J. & A. Cowles, to secure payment of two notes in the amount of $125 each.[37]

Not only was Frederick W. Crum's career in transition but he was also rather deeply in debt, when his wife, Ellice, died on May 16, 1846. She was buried in Unionville's Hillside Cemetery.[38] On October 23, 1846, Frederick W. Crum surrendered the parcel he purchased in 1844 to J. & A. Cowles in payment of the two outstanding notes, and also quitclaimed his life interest in his dwelling house to the Cowles brothers, conditioned on repayment of a note Crum had given them on November 23, 1836.[39] In addition, on October 23, 1846, Crum signed an agreement with Albert Hills to: "...board me in his family for five years to come to let and lease for the term of five years the dwelling house and other buildings connected therewith...containing about 1.5 acres of land belonging to Hellen Crum an heir of my late wife and in which I have a life interest being the same lately occupied by me with my family and now for a short time past occupied by the said Albert Hill..."[40]

Thereafter, Frederick W. Crum's fortunes improved. He and Albert Hill became partners. By 1850 they were manufacturing oyster tongs in the former Patent Wood Screw Manufactory building. Thereafter, and up until 1868, they manufactured crosscut saws.[41] By 1850 Crum had also married for a second time, to a woman named Susan, who died on April 13, 1902, at the age of 80. At some point Crum was able to clear his debts to J. & A. Cowles: he was still the owner of the same dwelling house when he died on June 1, 1895, at age 82, and was buried in Unionville.[42]

George Barber, Unionville

Little is known about George Barber's early background, other than that he was a Connecticut native, born about 1814, and arrived in Farmington at about the age of 20 in 1834. Owning no property, he was assessed only a poll tax of $20 on the town list of 1835. On April 6, 1835, Barber was admitted as a local elector. He was living in Avon, which abutted the village of Unionville on the north, at the time of his marriage to Delia Bull of Harwinton, on April 11, 1836.[43] However, Mrs. Delia (Bull) Barber died in 1839, at the age of 24.[44]

George Barber was likely an employee of the clock-making firm of Seymour, Hall & Co. as of

September 10, 1836, when Edward Seymour and Austin F. Williams deeded him a parcel of land in Unionville for only $1.00. The parcel, bounded north, east, and west by land of "the said Seymour and Williams," was undoubtedly the half-acre parcel (with no dwelling house) for which Barber was assessed taxes on the town's 1836 list. The probability that Barber was Edward Seymour's employee at the time is further suggested by the conditions of the transfer of the property, which were:

> ...if the grantee or his heirs, or the wife of said grantee, shall sell said premises to any other person...except the grantors or their heirs or in case the same should be taken on execution by a creditor of said grantor...he...shall pay to the grantors...the sum of One Hundred dollars, being the consideration for said land...[45]

So Barber didn't have to pay for the land, which was located a short distance southeast of Seymour and Williams's eastern clock factory building on Roaring Brook, unless he sold it to someone other than Seymour and Williams, or in the event it was taken on execution in a court case against them—an increasingly real possibility as time elapsed.

Meanwhile, as was the case elsewhere in the young United States during the late 1820s and early 1830s the unrestrained consumption of alcoholic beverages was becoming a problem in Unionville village. By 1830 Farmington's electors were pressuring the town selectmen to limit liquor licensing and retail sales,[46] and local records provide some indication of where many of the clockmakers stood on the issue. For example, in 1835 Williams, Orton, Prestons & Co.'s casemaking partner, Heman H. Orton, signed a highly controversial petition in favor of completely prohibiting the sale of spiritous liquors in Farmington.[47] The year 1842 witnessed the establishment of the Farmington Chapter of the Washington Temperance Society. Soon Austin F. Williams was elected chapter president.[48]

Sometime during 1835, a group of Unionville residents addressed the following "protest" "To the board, Civil Authority, Select Men, Constables, & Grandjurors of the Town of Farmington":

> Gent; Your petitioners would respectfully suggest to you, that a sense of duty impels us (however unpleasant the task may be,) to state to you, that the public house in this village [i.e.,

Unionville], kept by Gleason & Humphrey, is so managed & conducted, as to afford much facility for intemperate & Idle persons to mis[s]pend their time & money; that it is a common & almost constant occur[r]ence for a number of dis[s]olute & intemperate persons to assemble there & drink ardent spirits until they become intoxicated; and that in our opinion the establishment is a disgrace to the Town; & that it is doing incalculable mischief in this vicinity; That the proprietors have been repeatedly and urgently desired to correct their habits of dealing out spirits freely to the intemperate to no effect; & as a last resort we are constrained to represent their conduct to your Honorable board, - And pray you to not recommend them to the County Court as suitable persons to keep a House of public entertainment in future.[49]

George Barber's signature headed the list of petitioners. It was followed by the signatures of E.[dward N.] Saxton, Frederick W. Crum, No[a]diah Barber [unknown genealogical relationship to George Barber], Horace Wilcox, Asa B. Darrow, Benjamin C. Hosford, Daniel Bradley, Virgil Goodwin, Isaac W. Riggs, E.[dward] Seymour, A.[nson] Williams, D.[aniel] B. Johnson, Lynde Preston, James Roberts, D. Sperry, Jr., Wm. Griswold, John S. Hiscox, Sherman Pierpont, Oliver Moore, Lyman Boardman, Lowrey Hosford, Luther T. Parsons, Chauncey Douley[?], and Harvey Whittlesey. Several of these men have appeared previously, as partners, or as known or suspected employees, of Unionville's clock-making firms. The petition succeeded in its intended purpose. At a town meeting held in January 1836, Farmington's selectmen suspended all nominations of taverners for licensing. The process of nominating and granting taverners' licenses in Farmington was not resumed until 1841.[50]

By 1837 George Barber's local taxable estate had expanded to include both a watch and a dwelling house. A clock was included in his assessment for 1838, the same year in which the parcel of land he owned increased sharply in value—to $1,000. Evidently, Barber had improved it.

As explained in Chapter 3, on October 9, 1837, Seymour, Hall & Co. leased its entire clock-making facilities on Roaring Brook, which had recently been placed under attachment in several court cases, to Chauncey D. Cowles, son of the firm's financial backer, Timothy

Cowles.[51] Subsequently, in June 1838, Timothy Cowles acquired the property and buildings for unpaid debts upon execution of a court's judgment against Seymour and Williams. On February 15, 1839, George Barber deeded his nearby property with the buildings then standing thereon (now described as bounded on the north, east, and west on Timothy Cowles's), to William Bradley, Seymour, Hall & Co.'s successor, for $1,200.[52] Levi Smith and George Kilbourn, both connected with Unionville clock making, witnessed the deed. By July 15, 1839, George Barber had begun producing clocks under his own name. As we will see, perhaps no more than 250 of them were made.

Very few shelf clocks with labels identifying George Barber of Unionville as their makers have survived. Most have beveled cases. One such example is shown in Figures 100A and 100B. The large knob on its door may be a replacement. The clock once contained a 30-hr. wooden movement, now missing, and its source is uncertain. It is hoped that in the future, additional clock and movement data from intact examples will provide a key to a better understanding of Barber's clock-making activities at Unionville. The authors encourage anyone who encounters an intact example of a shelf clock with George Barber's Unionville label to contact the NAWCC Research Committee.

The label of the clock shown in Figure 100B contains the phrase "INVENTED BY ELI TERRY", a phrase consistently seen in labels of the clocks associated with the firms of Edward Seymour, further supporting an argument that George Barber had been connected with Seymour's firms. In addition, a State of Connecticut tax form, "for the year ending September 1, 1839," reported that 250 "Wooden Wheel Clocks" were in the process of being produced by George Barber in "Farmington" [Unionville] on July 15, 1839 (Figure 101).[53] This document not only provides firm evidence with which to date Barber's Unionville labeled clocks but also suggests the total number that were made.

George Barber produced clocks under his own name

Figure 100A. Clock case (movement missing), bearing label of George Barber, Unionville. Note the case's unusual beveled design.

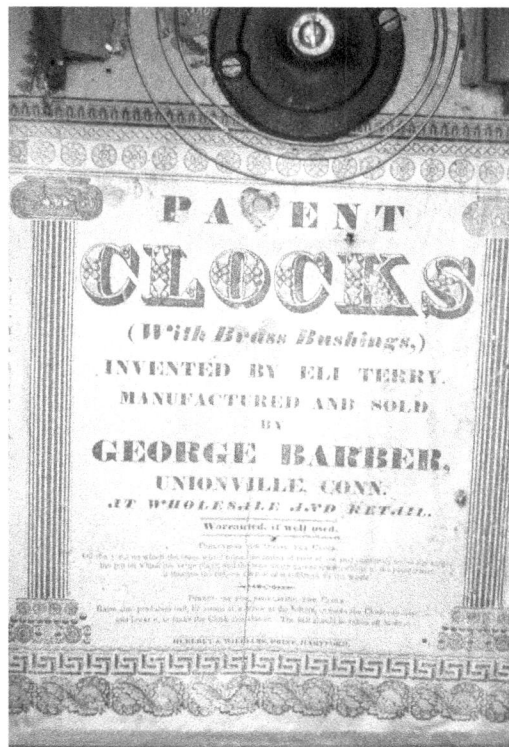

Figure 100B. George Barber, Unionville, label inside empty clock case of Figure 100A. The case once contained a 30-hr. wooden movement, now missing. (Figures 100A and 100B courtesy of Unionville Museum, MJD photos.)

STATE OF CONNECTICUT.

At a General Assembly of the State of Connecticut, holden at Hartford in said State, on the first Wednesday of May, in the year of our Lord one thousand eight hundred and thirty-nine.

SEC. 1. *Resolved by the Senate and House of Representatives in General Assembly convened,* That the several towns in this State, shall cause their Assessors to make a return to the Secretary of State, of the facts as they existed in each town, on the first day of September preceding, in relation to the number of kinds of manufacturing and mechanical establishments, the quantity and value of the articles manufactured, the number of persons employed, and the amount of capital invested; also in relation to the value of real estate, the number and value of horses, and other kinds of cattle; the quantity and value of the different products of land, and the number of persons employed in agricultural pursuits; the number, tonage, and value of the vessels of different kinds, and the nature and extent of the business in which they are engaged; and it shall be the duty of such Assessors to make said returns between the first day of November, and the first day of February next.

SEC. 2. The Secretary of State, shall cause to be printed, blank tables conveniently arranged for the return of the facts aforesaid, and shall furnish the Town Clerk of each town in the State, copies of the same, on or before the first day of October next.

SEC. 3. The Secretary of State, after he shall have received the returns aforesaid, from the Assessors of the several towns, shall cause to be prepared a true abstract of the same, and shall submit the same to the Legislature at the next Session thereof.

OFFICE OF THE SECRETARY OF THE STATE OF CONNECTICUT, JULY 15TH, A. D. 1839.—I hereby certify, that the foregoing Resolution is a true copy of record in this office. Attest, ROYAL R. HINMAN, *Secretary.*

STATEMENT *concerning certain branches of industry, as conducted by* George Barber *of* FARMINGTON, *for the year ending September 1st,* 1839, *agreeably to the directions of an Act passed May Session,* 1839.

Number of yards of Woolen Cloth manufactured in the family,	Number of pounds of Bar Soap manufactured,	
Gross value of Woolen Goods manufactured in the family,	Value of,	
The whole number of Saxony Sheep of different grades,	Number of Trunks made,	
The whole number of Merino Sheep of different grades,	Value of,	
The whole number of all other kinds of Sheep,	Number of hands employed,	
Whole number of pounds of Saxony Wool produced in said year,	Number of Sleighs made,	
Whole number of pounds of Merino Wool,	Fur blowing and Hat forming factories,	
Whole number of pounds of all other kinds of Wool,	Value of,	
Gross value of Wool produced in said year,	Capital invested,	
Amount of Capital invested in the growing of Wool,	Hands employed,	
Number of pairs of Boots manufactured during the year ending September first, one thousand eight hundred and thirty-nine and value of	Ready made Clothing,	
	Value of,	
Number of pairs of Shoes of all kinds and value,	Hands employed,	
Gross value of Boots and Shoes manufactured,	Number of Saddlery and Harness establishments,	
Number of Males employed in the business,	Value manufactured,	
Number of Females in same,	Number of hands employed,	
Number of Tanneries,	Number of pounds of Butter,	
Number of Hides of all kinds tanned during said year,	Value of,	
Gross value of Leather tanned and curried,	Number of pounds of Cheese,	
Number of hands employed,	Value of,	
Amount of Capital invested in the business,	Quarries of Stone,	
Number of Hat manufactories,	Value of manufactured,	
Number of Hats manufactured,	Hands employed,	
Gross value of Hats manufactured,	Number of barrels of Cider,	
Number of Males employed,	Number of Axe Helves manufactured,	
Number of Females,	Value of,	
Number of Paper Mills,	Hands employed,	
Number of Tons of Stock manufactured,	Number of Saw factories,	
Gross value of Paper manufactured,	Number of Saws made,	
Number of Males employed,	Value of,	
Number of Females,	Hands employed,	
Amount of Capital invested,	Number of bushels of Ruta Bags raised,	
Number of Axe manufactories,	Value of,	
Number of Axes manufactured during the year,	Number of bushels of Turnips,	
Gross value of Axes manufactured in same,	Number of thousand Teazles raised,	
Number of hands employed in the business,	Value of,	
Amount of Capital invested in same,	Number of pounds of Pork fatted,	
Number of manufactories of Cutlery,	Value of,	
Gross value of Cutlery manufactured in the year,	Number of Brass Bands, &c.	
Number of hands employed in the business,	Value of,	
Amount of Capital invested,	Number of Silver Plating establishments,	
Number of establishments for manufacture of Chairs and Cabinet Ware,	Value of Ware manufactured,	
Gross value of same manufactured in said year,	Hands employed,	
Number of hands employed in the business,	Number of Carriage Springs and Steps made,	
Number of manufactories of Tin Ware,	Value of,	
Gross value of Tin Ware manufactured in said year,	Hands employed,	
Number of hands employed in the business,	Number of machines for working Tin and Sheet Iron,	
Number of Distilleries,	Value of,	
Number of bushels of Grain distilled,	Number of hands employed,	
Number of gallons of Molasses distilled,	Number of Cords of Wood sold,	
Number of gallons of Spirits distilled,	Value of,	
Gross value of Spirit distilled,	Number of tons of Ship Timber,	
The number and value of Vessels—nature and extent of business—Sept. 1, 1839,	Value of,	
Amount of tonnage of the same,	Hands employed,	
Gross value of the same,	Number of Neck Stocks manufactured,	
Brass Wheel Clocks,	Value of,	
Wooden Wheel Clocks, 250	Hands employed,	
Locks for Doors and Cupboards,	Number of Window Sash and Door Blinds,	
Mouse and Rat Traps,	Value of,	
Chairs—Cabinet Ware,	Hands employed,	
Carriages—Waggons,		
Corn Brooms,		

Figure 101. State of Connecticut Comptroller's tax form, dated July 15, 1839, showing 250 "Wood Wheel Clocks" in progress by George Barber of Farmington [Unionville]. (Courtesy of Old Sturbridge Village microfilm collection, Connecticut State Library.)

for only a short time. Soon afterward, he apparently went to work for Williams, Orton, Prestons & Co., accompanying a load of the firm's clocks to Hartford on October 30, 1839, to be forwarded to New York City, and from there on to E[zra] N. Marvin, of Rockville [Parke County, IN].[54] Recalling that Barber's first wife died sometime in 1839, perhaps he accompanied the clocks to Indiana. In 1841 Barber reappeared on the town of Farmington's tax lists, when he was assessed only a poll tax. In 1842 the town levied "one assessment" of $20 on the firm of Crum & Barber.

About to depart from Connecticut for the American West, on July 1, 1843, Noah Preston, until recently a principal partner in Williams, Orton, Prestons & Co., deeded what had been his home, on a one-acre parcel in Unionville village, to Barber, for $1,400. The parcel was located in the vicinity of the one previously owned by George Barber, and near the two yellow factory boardinghouses then owned by Eli D. Preston and Cornelius R. Williams. The sale included the privilege of taking water from an aqueduct on Preston and Williams's land.[55] On October 14, 1843, George Barber mortgaged the property, together with an insurance policy covering the on-site buildings, to Elijah O. Gridley to secure a note.[56] Gridley quitclaimed his interest in the property back to Barber on September 6, 1844.[57] On August 2, 1848, George Barber, then residing in "Portage, Summit Co. Ohio," deeded the parcel and buildings to D.[avid] A. Keyes of Farmington,[58] terminating Barber's residence in Unionville—but not entirely severing his connection with clock making.

Returning to Frederick W. Crum's published recollections, after Crum & Barber abandoned their clock-making business, it was stated that George Barber moved to the West and "...succeeded financially in the furniture business."[59] Barber's withdrawal from the Unionville Ecclesiastical Society on June 1, 1844,[60] and the aforementioned deed of August 2, 1848, revealing that he had moved to Ohio, both support Crum's assertion to that effect. Indeed, a George W. Barber, age 36, described as a cabinetmaker born in Connecticut, was listed as a resident of "District No. 140," Portage, Summit Co., OH, in the 1850 U.S. Census. This George Barber had a wife named Sarah and two children born in Ohio circa 1844 and 1846, respectively. Also found in an 1859 Middlebury, Summit County, OH, city directory, was a listing for "George W. Barber & Co.," "furniture, ware rooms, & cain [sic] seat chair manufacturers," on "Howard St. north of Mill St."[61] Therefore,

it appears George Barber began using the middle initial of "W." upon moving to Ohio, to distinguish himself from another George Barber, also born in Connecticut, who already resided in the vicinity. The latter, a successful manufacturer of matches, was also found in the 1850 U.S. Census, also living in "District No. 140," but in Talmadge, Summit County, OH, with his wife and several children. The latter couple's children were all born in Ohio, circa 1839, 1841, 1843, and 1848, respectively. According to the census data, this man, born about 1806, was about eight years older than the George Barber who formerly made clocks at Unionville. Sometime after 1850, both Portage and Talmadge were absorbed into Middlebury, OH, which was in turn merged with Akron, OH, in 1872.[62]

Barber's involvement with clock making did not quite end with his move to Ohio. A very small number of beveled case shelf clocks with 30-hr. wooden movements, bearing the labels of "G.W. Barber, Middlebury, Ohio", attest to the fact that he cased a few wooden movements in Ohio. Figures 102A–102C illustrate one example. The label and movement of a second example are shown in Figures 103A and 103B.[63] Comparing Figures 102C and 103B, the movements in both of these clocks, although quite similar to one another, and also to the two versions of Williams, Orton, Prestons & Co.'s 30-hr. wooden movements described in Chapter 4, are attributed to Levi Smith, or one of Smith's firms. As mentioned above, Smith, who was a creditor of Williams, Orton, Prestons & Co.'s assigned estate in December 1841, is also thought to have provided movements for the clocks made by the firm of Crum & Barber. Note the positions of the count wheel retainer wires in the present examples, at 4:30 and 7:00 o'clock, respectively. The pendulum suspension studs in both examples are round, apparently a feature of Levi Smith's movements.[64]

George W. Barber spent the rest of his life in Akron, OH, and its environs. He died there, of tuberculosis, on June 12, 1875,[65] at the age of 62.

Kilbourn & Darrow, Unionville

A very small number of beveled case clocks are known that bear the labels of an obscure firm by the name of Kilbourn & Darrow of Unionville. An example is shown in Figures 104A–104C. As in the case of the relatively early clocks of Virgil C. Goodwin and the latter's clock-making firms, Kilbourn & Darrow apparently did not produce its own 30-hr. wooden movements, but instead purchased them from H. Welton & Co.[66] of Plymouth.[67] Although it is uncertain either exactly when Kilbourn & Darrow's enterprise commenced or when it ended, consistent with court record and label printer data, it was one of the last, if not the last, Unionville firm to produce wooden movement shelf clocks.[68] As shown below, the firm may have been in business as late as 1845.

George Kilbourn (1815–94) and Asa B. Darrow (1811–52), both of Farmington (Unionville), were the partners in Kilbourn & Darrow, per records of the execution of the Hartford County County Court's judgment in a case brought against the firm by Edward Phelps of Farmington.[69] On May 29, 1847, the official serving the execution stated in his return to the Hartford County County Court, that after visiting George Kilbourn, "one of the Debtors in this Ex[n] of money goods or Estate of the said Kilbourn to satisfy this Ex[n] and my fees but none was paid to me...and Asa B. Darrow being absent and out of the state therefrom... this Ex[n] remains wholly unsatisfied."[70] No other documents were found to explain what ultimately happened

in the matter of the unsatisfied execution.

Meanwhile, the partners in Kilbourn & Darrow became defendants in a second lawsuit. Each was "of Farmington" when summoned to appear before Roderick Stanley, Esq., a justice of the peace for Hartford County, on December 26, 1846. The purpose of the summons was to require the two men to answer charges brought by the firm of Whiting & Royce, consisting of partners Adna Whiting and John C. Royce, "manufacturers" of Farmington [specifically, Plainville village]. In their suit Whiting & Royce claimed that Kilbourn & Darrow had failed to pay a debt of $21.17 for 250 clock dials, a quantity of clock movement plates (apparently made of wood), and 167 feet of pine lumber, on Whiting & Royce's account books. The presence of clock plates in this account is interesting because it suggests Kilbourn & Darrow may have produced some of its own movements, albeit purchasing some parts thereof from Whiting & Royce (and perhaps others). Whiting & Royce was still in the clock dial and dial-painting business as late as 1850, at which time the firm employed six men. The court case records do not

Figure 102A. G. W. Barber, Middlebury, OH, wooden movement shelf clock in beveled case, exterior view.

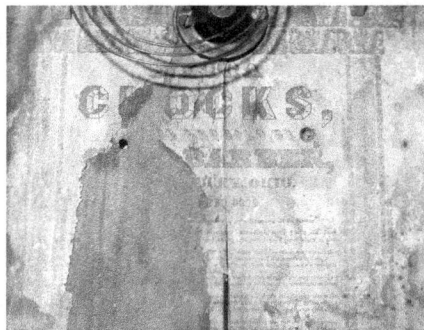

Figure 102B. G. W. Barber, Middlebury, OH, torn label inside clock shown in Figure 102A.

Figure 102C. Thirty-hr. wooden movement by Levi Smith or L. Smith & A. Blakeslee, Bristol, Type 7.154 (Snowden Taylor, unpublished), inside G. W. Barber clock shown in Figures 102A and 102B. (Figures 102A–102C courtesy of Joe M. Devilbiss.)

identify either the date on which Kilbourn & Darrow incurred the debt to Whiting & Royce, or the case's outcome.[71]

As mentioned in Chapter 6, George Kilbourn was born in Litchfield, CT, on December 27, 1815, to parents Norman and Lucy (Peck) Kilbourn. In October 1835 he testified as a witness in the suit brought by the young Virgil C. Goodwin against clock factory owner Henry Hart of Goshen. This information suggests that Kilbourn had been working with or for Goodwin at Goshen in 1832–34, to produce the batch of about 900 wooden shelf clock movements that became the subject of Goodwin's suit. Kilbourn and Goodwin evidently became friends, both moving to Unionville at about the time Goodwin's case against Hart commenced in October 1834. Together with Russel Humphrey, Goodwin and Kilbourn were convicted in November 1835 of assaulting John S. Hiscox, the latter a "mechanic"—and probably, like his assailants, a clock factory employee. The incident took place in Unionville.[72]

George Kilbourn first appeared on the town of Farmington tax lists in 1835, when he was assessed for personal property consisting solely of one watch. Undoubtedly, he was a boarder at the time. On July 4, 1839, George Kilbourn "of Litchfield, now of Unionville" married Betsey Wright of Hartland (CT). The wedding took place in Farmington.[73] In 1841 Kilbourn was appointed the first clerk of the newly formed Unionville Ecclesiastical Society.[74] He was assessed a $20 poll tax on Farmington's list of 1842 and appeared on the town's tax lists during succeeding years through 1844.

By 1843 George Kilbourn was involved with Virgil C. Goodwin and Russel Humphrey (recent copartners in the clock casing or sales firm of Goodwin &

Figure 103A. Label inside a second example of a 30-hr. wood movement shelf clock by G. W. Barber, Middlebury, Ohio.

Figure 103B. Movement by Levi Smith or L. Smith & A. Blakeslee, in G.W. Barber, Middlebury, OH, shelf clock shown in Figure 103A. (Figures 103A and 103B Ward Francillon photos.)

Figure 104A. Thirty-hr. wooden movement shelf clock by Kilbourn & Darrow, Unionville, exterior view.

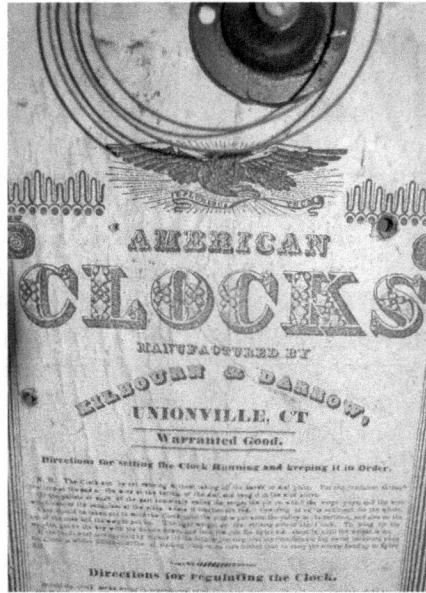

Figure 104B. Kilbourn & Darrow, Unionville, label inside clock shown in Figure 104A.

Figure 104C. Thirty-hr. wooden movement by H. Welton & Co. inside Kilbourn & Darrow clock shown in Figures 104A and 104B. (Figures 104A–104C, courtesy private collection, MJD photos.)

Humphrey), in a short-lived business venture, when the extant accounts of Unionville farmer and blacksmith George Payne named him as a partner in the firm of "Kilbourn & Humphrey," on February 20. About the same time, Kilbourn & Humphrey was credited "by an order on Noah Preston to be paid in wood." On April 5, 1843, however, under Kilbourn & Humphrey's account, a debit to "George Kilbourn & [Virgil C.] Goodwin" was recorded for an unspecified amount of cash and merchandise. The last entry for the firm of Kilbourn & Humphrey, dated October 10, 1845, indicated that the account had been settled.[75]

On December 26, 1844, Daniel B. Johnson deeded to George Kilbourn a one and a half acre parcel of land situated "near Unionville on the road to Canton through Lovely Street," bounded north by William Griswold, east by a highway, south by land owned by the grantor, and west by J. & A. Cowles and John Thompson.[76] The parcel lay in close proximity to the former Patent Wood Screw Manufactory in which the clock-making firm of Williams, Orton, Prestons & Co. had operated until its dissolution early in 1842. Nevertheless, the building continued to be designated as "the clock factory" on the Town of Farmington's tax

lists through the year 1847, so presumably, Kilbourn & Darrow was one of the last firms to assemble clocks there. In Unionville's close-knit community, the property purchased by Kilbourn also lay in proximity to the dwelling houses of the then present or former clockmakers Frederick W. Crum, Virgil C. Goodwin, and Asa B. Darrow. In 1845 Kilbourn was assessed taxes on the parcel and also on one horse, one "neat cattle," one poll, and one clock. Kilbourn withdrew from the Unionville Ecclesiastical Society in 1846, removing with his family to New Britain sometime afterward.[77]

With his removal from the village, George Kilbourn's involvement with clock making ended, at least for a time. The 1850 U.S. Census found him working in New Britain as a clerk. Kilbourn's longtime friend, clockmaker Virgil C. Goodwin, appeared on Farmington's tax list for 1846, but afterward removed with his family to join the Kilbourns at New Britain, where according to the 1850 U.S. Census, Goodwin found work as a locksmith. By 1860 both men had moved again, this time to New York State, where Virgil Goodwin sold sewing machines in Rochester, and George Kilbourn, then a resident of Portville, Chattaraugus Co., NY (on the Allegheny River near Olean), once again found employment as a clockmaker! Later, Kilbourn returned to Connecticut, where he died on November 1, 1894, at the age of 79, and was buried in his hometown of Litchfield.[78]

Asa B. Darrow, George Kilbourn's partner in the firm of Kilbourn & Darrow, was born in Connecticut circa 1811, the son of Asa Darrow Jr. (1783–1848) of Farmington. On October 14, 1834, Asa B. Darrow "of Farmington" married Sarah Garner of Goshen.[79] During May and June 1839 Asa and Sarah lost two young children; both were buried in Avon's Greenwood Cemetery. Sarah Darrow died, and Asa remarried, sometime prior to 1850.

We have encountered Asa B. Darrow previously; for example, in Chapter 2 he appeared in the account books of the clock-making firm of Seymour, Williams & Porter. In the accounts, Seymour, Williams & Porter paid Darrow for unspecified services beginning in April 1835. Darrow was also a codefendant, together with Horace Wilcox, George Barber, Anson Williams, Edward Sexton, Daniel B. Johnson, and (probably) Frederick W. Crum, when the men were accused of assaulting Levi B. Rowe in January 1835, just days before the dissolution of the firm of Seymour, Williams & Porter. Two of Darrow's codefendants in the case, Anson Williams and Edward N. Sexton, were silent partners in the clock-making firm of Seymour, Hall & Co., the successor to Seymour, Williams & Porter. Asa B. Darrow's name began appearing on Farmington's tax lists in 1838 and continued to appear thereon as late as 1846.

Aside from his role as a partner in the firm of Kilbourn & Darrow, there is more to the story of Asa B. Darrow's connection with clock making. On November 22, 1843, a sheriff's deputy of Hartford County summoned him to appear before the Hartford County (CT) Superior Court on the "fifth Tuesday of January" 1844, to answer a suit brought by clockmaker Chauncey Jerome, who was then of Bristol.[80] Briefly, Darrow and his codefendant in the suit, Ransom Blakeslee Jr., had given Jerome a note on April 22, 1843, "jointly and severally" promising to pay him $518 with interest, "for value received," perhaps referring to clocks or movements. The sheriff's deputy attached as Asa B. Darrow's property a piece of land in Unionville, formerly comprising Lot #10 in the original subdivision of Solomon Langdon's farm, containing "ten acres more or less with a dwelling house and barn standing thereon, bounded

Figure 105A. Samuel E. [?] Curtiss, Unionville shelf clock, exterior view, showing beveled case.

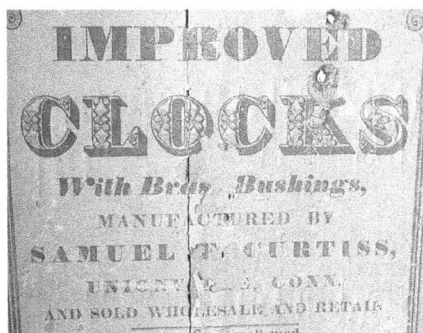

Figure 105B. Label of Samuel E. [?] Curtiss, Unionville, shelf clock of Figure 105A.

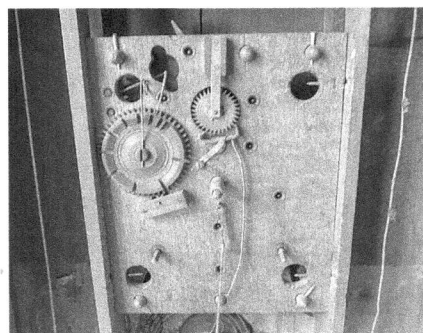

Figure 105C. Thirty-hr. wooden movement by H. Welton & Co., inside Samuel E. [?] Curtiss shelf clock shown in Figures 105A and 105B. The movement's verge retainer is a replacement. (Figures 105A–105C courtesy of Cliff Alderman.)

north on Luther Parsons, east on Timothy Cowles, south on Jeptha Hart, and west on highway, being the same premises now occupied by the Family of the Defendant..."[81] Darrow had purchased the parcel from Daniel Woodruff for $130 in December 1838. By 1840 a house stood on the property.[82]

Jerome's court case against the two men raises the possibility that Asa B. Darrow and Ransom Blakeslee Jr. (1814–?) had been involved in a clock sales business in 1843 or earlier. It is interesting that on April 2, 1844, Darrow and Blakeslee (sometimes spelled "Blakesley") arrived together in New York City after a trip to London. Upon his arrival Darrow's occupation was described as "joiner."[83] Perhaps it is not a coincidence that about 1844, Chauncey Jerome began selling his 30-hr. brass movement shelf clocks at a profit in England,[84] although the authors do not know the purpose of Darrow and Blakeslee's trip, or whether it might have been connected with Jerome. At the time of Jerome's suit, Blakeslee was a resident of Plymouth (CT). Between 1846 and 1851 Ransom Blakeslee Jr. resided in New York City, selling clocks manufactured at Plymouth under the firm name of R. Blakeslee, Jr. & Co.[85] The partner(s) comprising the "& Co." in the latter firm are unknown.

Darrow and Blakeslee defaulted on their scheduled court appearance. Consequently, the court issued execution of its judgment in favor of Chauncey Jerome on October 3, 1844. Briefly, with the help of two men, Phineas Curtis and Joshua Moses, who held mortgages on the aforementioned land and buildings,[86] Asa B. Darrow was able to pay the debt to Jerome soon after the execution. Jerome quitclaimed his interest in the land and buildings back to Darrow on February 22, 1845.[87] Asa B. Darrow was able to clear the mortgages, selling the parcel to Albert Hill in 1847.[88]

The 1850 U.S. Census found Asa B. Darrow living in Chenango, Broome County, NY, with a new wife, Hannah, and a nine-year-old son born in New York State about 1841. Darrow died and was buried at Binghamton, Broome County, NY, a few days before August 24, 1852, at the age of 41.[89]

Chapter 2 mentioned a connection between Timothy Porter, a prominent joiner of Farmington village who had been a partner in the clock-making firm of Seymour, Williams & Porter, and the construction of several public buildings in Macon, GA. A small number of shelf clocks that bear the labels of a firm known only as "Darrow & Seymour" of Macon, GA, have been

reported. Movements by various makers other than those of Farmington and Unionville have been observed in the mysterious Darrow & Seymour labeled clocks. At least one example, with an 8-day brass, rack and snail weight movement made by the firm of C.[hauncey] & N.[oble] Jerome, with a Darrow & Seymour label pasted over the original C. & N. Jerome label, has been reported,[90] and it is known that Edward Seymour's firms sold clocks in Alabama and Georgia. All this fuels a speculation that the "Darrow" in Darrow & Seymour might have been Asa B. Darrow.

Samuel E.[?] Curtiss, Unionville

A single beveled case clock bearing the label of Samuel E. [?] Curtiss, Unionville, has been reported (Figures 105A–105C).[91] Its brass bushed, 30-hr. wooden movement (Figure 105C), with a replaced verge retainer, is attributed to H. Welton & Co. of Plymouth.[92] Dating from approximately 1839 or the early 1840s, the clock and its (mostly legible) label (Figure 105B), seems to be connected with Samuel E. Curtiss (1808–83), a shoemaker of Southington and Berlin, CT, and Broadalbin, NY, who became a professional photographer late in life.[93] It is unclear whether this was the same Samuel Curtis[s] [no middle initial], who appeared on the Town of Farmington's tax lists for the years 1833–34, and 1839–42. In any event it is possible that Samuel E. Curtiss obtained a few clocks from Virgil C. Goodwin or one of Goodwin's firms (who also used 30-hr. wooden movements obtained from H. Welton & Co.), in trade, placing his own labels inside their cases.

As a manufacturer of shoes, Samuel E. Curtiss shared a common interest with a number of Connecticut's clockmakers and their financiers, including Edward Seymour, Frederick P. Hall, Austin F. Williams, Cornelius R. Williams, and Timothy Cowles, in recognizing the need for a national bank—a highly controversial and politically charged matter at the time. Indeed, all of the men's signatures appeared on a petition dated June 26, 1834, addressed to the United States House of Representatives (published as Document No. 508 in: *House Documents, 23rd Congress, 1st Session, 1833–1834*), expressing support for the United States Bank. Nonetheless, the authors do not know whether Curtiss might have provided the boots sold by the firm of Gillet & Hall, along with clocks made by Seymour, Hall & Co., in the states of Alabama and Georgia, between 1835 and 1837.

Samuel E. Curtiss, then "of Southington" married Mary Bidwell Andrews (1807–?), the daughter of

Ezekiel and Ruth (Bidwell) Andrews of Berlin, on April 21, 1830.[94] Between March 1829 and September 1837 Curtiss acquired several properties in Southington, some located along the Meriden-Waterbury turnpike, and one bounded north and west by the Tenmile River,[95] which he later sold. By July 4, 1845, Samuel and his wife had removed from Connecticut and were living in Broadalbin, Fulton County, NY.[96] Nothing further is known about Samuel E. Curtiss or his mysterious connection with Unionville.

A Note on Levi Smith of Farmington and Bristol

Although his name has not been found on the labels of any shelf clocks produced in either Farmington or Unionville villages, a few words must be said about Levi Smith (ca. 1807–80), who apparently sold or traded 30-hr. wooden movements to several Unionville clockmakers and firms. Very little is known about this man, undoubtedly the "L. Smith" who, along with Frederick W. Crum and George Barber, appeared in the early accounts of the clock-making firm of Seymour, Williams & Porter. During the 1830s Smith also served as a justice of the peace for the town of Farmington.

Levi Smith arrived in Farmington shortly before April 1, 1833, on which date he was admitted as an elector "By Certificate...from Plymouth." It seems significant that several other individuals soon to play prominent roles in the history of clock making in the villages of Farmington and Unionville were also admitted as electors in the town of Farmington on the same day: Cornelius R. Williams, Henry Tolles, and William L. Gilbert, all "by Certificate" from Bristol; George Marsh "by Certificate" from Litchfield; and Isaac P. Frisbie "by Certificate" from Windsor.[97] Thus, it appears that the men arrived as a group, having previously worked together in Marsh & Gilbert's clock factory at Bristol, to work for George Marsh's clock-making firms of Farmington village and their successors.

Levi Smith was described as being "of Northfield" when he married Lucy Ann Bancroft on March 25, 1834.[98] Lucy Ann was born on March 10, 1814, to parents Luman and Clarisa Bancroft of Harwinton. Although it is uncertain, Lucy Ann (Bancroft) Smith may have been the sister of Henry Bancroft, who married a younger sister of Eli D. and Noah Preston, Amelia [sometimes "Aurelia"] Preston, of Harwinton, on September 6, 1836.[99] Henry Bancroft's name was mentioned in Chapter 4, in connection with a boarding house associated with the former Patent Wood Screw Manufactory, beginning in 1841 and extending through 1847, while the building was being used as a clock factory.[100]

On February 6, 1838, Levi Smith purchased a parcel of land in Unionville, bounded east on "land of Edward Seymour, South on highway, West on land of Daniel Woodruff, North on land formerly owned by [Solomon] Langdon deceased, and contains six acres..." from William Mather of Simsbury and Charles T. Hillyer of Granby.[101] On the same day, Mather and Hillyer deeded an additional eight-acre parcel located in Unionville to Smith, bounded east and west by land of Edward Seymour and others, south by the Farmington River, and north by land of Carrington and Wilcox, with a highway running through it. Smith sold the latter parcel to Gideon Walker (whose lands abutted those of Williams, Orton, Prestons & Co. in Unionville), on April 5, 1838, but reserved a half-acre portion thereof.[102] On July 12, 1839, Levi Smith quitclaimed the remaining half-acre parcel, now described as "a portion of Lot No. 3 in the late Solomon Langdon's subdivided estate," to Unionville clockmaker Virgil C. Goodwin.[103] This was the southernmost portion of the former Lot No. 3, bounded south on the Farmington River.

Shortly after quitclaiming the parcel to Goodwin, Smith removed to Bristol. In 1840, together with Augustus Blakeslee and Edwin Ray under the firm name of Smith, Blakeslee & Co., Levi Smith commenced producing 30-hr. wooden clock movements, continuing the business under his own name beginning about 1842. At least two other individuals formerly associated with Edward Seymour's firms in Unionville, George Barber and Virgil C. Goodwin, commenced assembling or casing clocks under their owns names in 1839, about the same time Smith moved from Farmington to nearby Bristol. In December 1841 Levi Smith's name was listed among those of the approved creditors of Williams, Orton, Prestons & Co.'s assigned estate, suggesting that Smith had provided movements or movement parts to the firm as it worked to produce clocks from its unfinished stock. Smith's movements are found in the clocks of Crum & Barber. All this is particularly interesting in light of Frederick W. Crum's published recollection that Crum & Barber purchased a portion of Williams, Orton, Prestons & Co.'s business in 1841.

Levi Smith was still living in Bristol, when, on March 19, 1844, Eli Dewey Preston, one of the principal partners in the former firm of Williams, Orton, Prestons &

Co., initiated a lawsuit against him. In his suit Preston charged Smith with failing to pay a promissory note he had given Preston on August 30, 1843, in the amount of $262.63, payable "On demand for value received."[104] What "value" Smith received of Preston was not stated in the documents examined, and there is no indication that clock parts or machinery changed hands. On March 25 Bristol's deputy sheriff Asa Bartholomew went out and attached four horses and four wagons as Smith's property. The case continued until the January 1846 term of Hartford County Superior Court, when Smith defaulted on his appearance. Subsequently, the Court rendered its judgment in favor of Preston, with "Exn [Execution] issued March 4, 1846."[105]

Based on probate records of Levi Smith's 1845 insolvency, in addition to wooden movement making, Levi Smith was operating a livery stable in Bristol.[106] After the insolvency proceedings ended, he continued to produce wooden clock movements as late as 1847 and was one of Connecticut's last producers thereof.[107] At the time of the 1850 U.S. Census, Smith's occupation was still "Clocks." In 1860 his occupation was that of "mechanic." The years 1870 and 1880 found him working in an unidentified clock factory at Bristol.

Levi Smith died at Bristol on June 28, 1880. His widow, Lucy Ann Smith, died there in 1884. Both are buried in Bristol's West Cemetery.[108]

Conclusions

The authors have presented much new information relative to the clock-making firms of Farmington and Unionville villages, their partners, and their chronologies. Two distinct clock-making "shops" or "branches" in these villages have been identified; the first was the firms of Edward Seymour and their successors. The second, composed of George Marsh's firms and their successors, was somewhat larger and more enduring. Both converged around a single funding source: Farmington merchant Timothy Cowles. As each of the branches faded, clocks were assembled from movements and parts exchanged with others, and the products of the respective branches lost much of their distinctiveness.

We have glimpsed how many of the challenges and controversies faced by the young American nation in the decades leading up to the Civil War, for example, also touched the clockmakers' lives (e.g., industrialization, technological change, slavery and abolitionism, child labor, banking, westward migration, temperance, the speculative boom leading up to the Panic of 1837, and the ensuing economic depression).

In May 1845 the Office of Connecticut's Secretary of State issued a report on the status of manufacturing in the various towns, helping to answer the question of when clock making ended in Unionville village. According to the report, in 1845 a total of four clock manufactories were operating in the town Farmington, producing a total of 9,500 clocks per annum. The combined value of the manufactories' output was $19,875, and they employed a total of 18 people.[109] As we have seen, by 1845, clock making in Farmington and Unionville villages, which consisted predominantly of the wooden movement variety, had largely been abandoned. The firm of Kilbourn & Darrow was probably one of the holdouts included in the statistic. Virgil C. Goodwin may have been another.[110]

Although our discussion has of necessity been limited to clock making in the villages of Farmington and Unionville, during the course of this study the authors became aware that clock casemaker John Hunt of Plainville village (also within the town of Farmington) was still producing shelf clocks with movements obtained from others as late as August 21, 1846, when a batch of clocks in progress at his shop was placed under attachment in a court case.[111] The brass movement clock-making firm of Hills, Goodrich & Co., together with clockmaker Joseph Ives, at Plainville village,[112] probably accounted for the remaining clock-making firm in the Secretary of State's 1845 report relative to the town of Farmington.

During this study the authors became aware of several connections between the clockmakers of Farmington's three villages, which may serve as a foundation for further research relative to early nineteenth-century shelf clock making at Plainville village. For example, at least one clockmaker of Farmington village, Henry Tolles, a partner in Orton, Preston & Co. who left the firm in 1835, later worked at Plainville for Hills, Goodrich & Co.[113] A number of lawsuits involving casemaker John Hunt of Plainville village intersected Unionville's clock-making sphere. Austin F. Williams, son-in-law of Timothy Cowles, the financier of several of Unionville's clock-making firms, sold lumber from his yard in Plainville village to a number of Connecticut's clockmakers. Brass movement making was carried on within a portion of Williams's lumberyard prior to 1836 (see footnote 53, Chapter 4). In addition, Austin F. Williams owned a depot on the Farmington Canal at "Bristol Basin" in Plainville village that served as a central agency for the shipment of clocks and clock-making supplies prior to the Canal's closure in 1848, following closely the end of clock making in Unionville.

Finally, it should be noted that, subsequent to the firm of Hills, Goodrich & Co., George Hills was active in brass movement clock making under his own name in Plainville village between 1850 and 1870.[114] Thus, as of 1850 and beyond, a few Farmington residents still listed their occupations in the U.S. censuses as "clock maker."

Notes and References

1. Kenneth D. Roberts and Snowden Taylor. *Eli Terry and the Connecticut shelf clock*. 2nd ed. Fitzwilliam, NH: Ken Roberts Publishing Co., 1994.

2. See, for example, Chapter 5, for sale price data for clocks sold at the 1845 auction of Noah Preston's Illinois estate.

3. Roberts and Taylor, *Eli Terry and the Connecticut shelf clock*, 1994.

4. According to city directory data compiled by D. R. Slaght (unpublished, NAWCC Library), printer Joseph Hurlbut of Hartford, who supplied labels for Crum & Barber's clocks, was at work from 1832 to 1838 and again from 1840 to 1844. In 1839 he was a copartner in the printing firm of Hurlbut & Williams. Furthermore, Crum & Barber's beveled clock cases suggest a date no earlier than 1839.

5. Town of Farmington, Tax Records, Box 12 (1837–42), RG62, Connecticut State Library.

6. Vital Records Index, Connecticut State Library.

7. Spencer B. Reynolds, comp. *Descendants of the immigrant Thomas Catlin*. Harwinton, CT: East Academy, 1985.

8. Farmington Land Records (FLR) Volume (V.) 46, p. 398, May 2, 1836, recorded December 7, 1836.

9. Ibid.

10. FLR V. 49, p. 406, December 7, 1836.

11. FLR V. 48, p. 298, May 4, 1840; see also FLR V. 47, p. 32, November 2, 1833, for a description of the location of the former Patent Wood Screw Manufactory.

12. *Bristol Herald*, July 3, 1890.

13. *Connecticut Comptroller's Return Schedules, Town of Farmington, Ct. 1839*. Old Sturbridge Village collection on microfilm, Connecticut State Library.

14. Farmington Town Records, V. 2, Part 2, Farmington Tax Assessor's Office.

15. Austin F. Williams papers, Connecticut Historical Society.

16. FLR V. 46, p. 441, dated September 10, 1836, and April 3, 1837. It appears grantors Seymour and Williams signed the document on different dates, nearly seven months apart.

17. FLR V. 46, p. 573, February 15, 1839.

18. *Connecticut Comptroller's Return Schedules, Town of Farmington, Ct. 1839*.

19. *Connecticut Courant*, February 4, 1837.

20. *Connecticut Courant*, August 18, 1837.

21. Noah Preston Probate File, Illinois Regional Archives Depository, Southern Illinois University, Carbondale, IL.

22. Barber was not found on the Town of Farmington Assessor's lists for either the years 1839 or 1840.

23. FLR V. 48, p. 433, July 1, 1843.

24. FLR V. 49, p. 424, August 2, 1848.

25. See Chapter 4 for a discussion of Williams, Orton, Prestons & Co.'s assigned estate.

26. Farmington Probate District Estate Papers for Williams, Orton, Prestons & Co. Connecticut State Library.

27. See *Connecticut Courant*, May 15, 1841.

28. Snowden Taylor (unpublished manuscript); and Roberts and Taylor, *Eli Terry and the Connecticut shelf clock*, 1994.

29. Town of Farmington Records, Vol. 2, Part 2, Farmington Tax Assessor's Office.

30. It is interesting that based on unpublished information in author ST's research and photo files, the two basic types of 30-hr. wooden movements attributed to Levi Smith or Smith's firms are barely distinguishable from the two basic types attributed to Williams, Orton, Prestons & Co. Key distinctions center on a round pendulum suspension stud (Smith) vs. a rectangular stud (W.O.P. & Co.). Variability in the shape of movement pillar posts (possibly sourced from outside the respective clock factories) has also been observed.

31. Accounts of George Payne 1842–96. Farmington Public Library.

32. At the time of the 1850 U.S. Census (Farmington, CT), Eber N. Gibbs was a clockmaker.

33. FLR V. 47, p. 366, February 27, 1843.

34. FLR V. 48, p. 536, April 29, 1844.

35. FLR V. 48, p. 537, June 14, 1844.

36. FLR V. 45, p. 536, January 31, 1845; FLR V. 48, p. 566, January 31, 1845.

37. FLR V. 49, p. 57, August 16, 1845.

38. Ancestry.com (online source).

39. FLR V. 50, pp. 75–76, October 23, 1846.

40. FLR V. 49, p. 586, October 23, 1846.

41. Christopher P. Bickford. *Farmington in Connecticut*. 2nd ed. Farmington, CT: Farmington Historical Society, 2008.

42. Death Records Index, Connecticut State Library.

43. Barber Collection, unpublished photocopy of Connecticut Vital Records, V. 54, Town of Harwinton, Connecticut State Library. Delia Bull (who apparently had a brother named Henry) does not appear to have been a close relative of Henry C. Bull, the partner in Williams, Orton, Prestons & Co.

44. Ancestry.com (online source).

45. FLR V 46, p. 441, September 10, 1836, and April 3, 1837. (See also Note 16 above.)

46. Christopher P. Bickford. *Farmington in Connecticut*, 2008.

47. John Treadwell papers (1706–1872), Box 1, "Miscellaneous Papers," RG69:25, Connecticut State Library.

48. Washington Temperance Society of Farmington Record Book 1842–54 (photocopy), Connecticut Historical Society.

49. Town of Farmington, Administrative Records—Petitions, RG:62-052, Connecticut State Library.

50. Mabel S. Hurlburt. *Farmington town clerks and their times*. Hartford, CT: Press of Finley brothers, 1943.

51. FLR V. 47, p. 229, October 9, 1837.

52. FLR V. 46, p. 573, February 15, 1839.

53. *Connecticut Comptroller's Return Schedules, Town of Farmington, Ct. 1839*.

54. Noah Preston Probate File, Illinois Depository.

55. FLR V. 48, p. 433, July 1, 1843.

56. FLR V. 48, p. 435, October 14, 1843.

57. FLR V. 45, p. 519, September 6, 1844.

58. FLR V. 49, p. 424, August 2, 1848.

59. *Bristol Herald*, July 3, 1890.

60. First Church of Christ, Unionville Records from 1839, Connecticut State Library.

61. C. S. Williams. *Williams' Akron, Wooster, and Cuyahoga Falls [Ohio] Directory, City Guide and Business Mirror*, Vol. 1, 1859–60. Akron, OH: W. G. Robinson, Wooster & J. H. Baumgardener & Co., 1859.

62. See, for example, Wikipedia (online source).

63. "The Answer Box." *NAWCC Bulletin*, No. 97 (April 1962): 249.

64. Snowden Taylor, Unpublished manuscript.

65. Ancestry.com (online source).

66. Snowden Taylor. "Characteristics of standard Terry-type 30-hr. wooden movements as a guide to identification of movement makers." *NAWCC Bulletin*, No. 208 (October 1980), and Snowden Taylor. "More on H. Welton & Co. wood movements." *NAWCC Bulletin*, No. 227 (December 1983): 737–739.

67. Roberts and Taylor, *Eli Terry and the Connecticut shelf clock*, 1994.

68. Ibid.

69. *Edward Phelps vs. Kilbourn & Darrow*. Hartford County County Court, Executions, Box 533, RG3, Connecticut State Library.

70. Ibid.

71. 1850 U.S. Census of Manufactures; and *Whiting & Royce vs. Kilbourn & Darrow*. Town of Farmington, Justice of the Peace Records, Box 41, RG62, Connecticut State Library.

72. *State vs. Russel Humphrey et al.* Hartford County County Court Files, Box 427, RG3, Connecticut State Library.

73. Vital Records Index, Connecticut State Library.

74. David D. March. "Semi-centennial anniversary of the First Church of Christ, Unionville, 1841–1891." Connecticut State Library.

75. Accounts of George Payne 1842–96. Farmington Public Library.

76. FLR V. 48, p. 557, December 26, 1844.

77. First Church of Christ, Unionville Records from 1839, Connecticut State Library.

78. Vital Records Index, Connecticut State Library.

79. Vital Records and Marriage Indexes, Connecticut State Library.

80. *Jerome vs. Darrow and Blakeslee*. Hartford County County Court Executions, Drawer 101, RG3, Connecticut State Library.

81. See FLR V. 48, p. 280, September 3, 1841; FLR V. 47, p. 408, October 3, 1844; and FLR V. 45, p. 537, February 22, 1845.

82. FLR V. 46, p. 559, December 1838. See also Betty Coykendahl. "History of Unionville Houses." Unpublished manuscript, 1979, Farmington Public Library.

83. New York Passenger and Immigration Index and Lists (microfilm), Connecticut State Library.

84. Chris H. Bailey. "From rags to riches to rags: the story of Chauncey Jerome." *NAWCC Bulletin* Supplement No. 15 (Spring 1986): 65–68.

85. Sonya L. Spittler, Thomas J. Spittler, and Chris H. Bailey. *Clockmakers and watchmakers of America by name and by place.* 2nd ed. Columbia, PA: National Association of Watch & Clock Collectors, 2011.

86. FLR V. 48, p. 166, February 20, 1841; and FLR V. 48, 391, April 1843. See also Betty Coykendahl, "History of Unionville Houses," 1979.

87. FLR V. 45, p. 537, February 22, 1845.

88. Betty Coykendahl, "History of Unionville Houses," 1979.

89. Death Index, Connecticut State Library; and *Hartford Courant*, August 28, 1852.

90. "Research Activities & News." *NAWCC Bulletin*, No. 261 (August 1989):345.

91. Cliff Alderman, personal communication to MJD, July 2013.

92. Snowden Taylor, Unpublished manuscript. The movement is classified as subtype 8.121.

93. Frederic H. Curtiss. *A genealogy of the Curtiss family.* Boston, MA: Rockwell and Churchill Press, 1903. Based on the information provided in this source and in *A genealogy of the Curtiss-Curtis family of Stratford, Connecticut,* compiled by Harlow D. Curtis (Ann Arbor, MI: Edwards Bros., 1953); and also in *A family named Curtis,* compiled by Rose Mary Goodwin (Privately published, 1983), it appears unlikely that the name appearing on the clock's label was that of Samuel F.[osdick] Curtis[s] (1799–1871), a native of Wethersfield, CT, who removed to Pen Yan, Yates County, NY, about 1824, evidently spending the rest of his life there. A staunch abolitionist, Samuel F. Curtis was elected president of Pen Yan village in 1830. The 1850 U.S. Census lists his occupation as chair maker.

94. Vital Records Index, Connecticut State Library.

95. See, for example, the following Town of Southington Land Records: V. 13, p. 101, March 6, 1829; V. 13, p. 293. October 8, 1829; V. 13, p. 384, March 6, 1829; V. 14, p. 420, October 23, 1832; and V. 15, pp. 616–617, September 18, 1837.

96. Southington Land Records, V. 17, pp. 575–576, July 4, 1845.

97. Town of Farmington Records, Vol. 2, Part 2, Farmington Tax Assessor's Office.

98. Harwinton Vital Records, Connecticut State Library.

99. Vital Records Index, and Harwinton Vital Records, Connecticut State Library.

100. Town of Farmington Tax Records (various boxes), RG62, Connecticut State Library.

101. FLR V. 46, p. 507, February 6, 1838.

102. FLR V. 46, p. 529, February 6, 1838; and FLR V. 48, p. 8, April 5, 1839.

103. FLR V. 45, p. 397, July 12, 1839.

104. *Eli D. Preston vs. Levi Smith*. Hartford County Superior Court Files, Box 103, RG3, Connecticut State Library.

105. Ibid.

106. Bristol Probate District Estate Packets, Connecticut State Library.

107. Roberts and Taylor, *Eli Terry and the Connecticut shelf clock,* 1994.

108. Clifford Rourke (compiler). "Bristol [CT] Headstone Inscriptions." Unpublished, 1934, Connecticut State Library.

109. Connecticut Secretary of State. *Report of the Connecticut secretary of state relative to certain branches of industry, May Session 1845*. Hartford, CT, 1845.

110. See Chapter 6.

111. *Wells, Hendrick & Co. vs. John Hunt*. Hartford County County Court Files, Box 453, RG3, Connecticut State Library.

112. Kenneth D. Roberts. *The contributions of Joseph Ives to Connecticut clock technology* 1810–1862. 2nd ed. Fitzwilliam, NH: Ken Roberts Publishing Co., 1988.

113. Jacque Houser. "A close look at Hills, Goodrich & Company, its partners, its products, its employees, and its relationship with Joseph Ives." *Timepiece Journal of the American Clock & Watch Museum*, Vol. 5, No. 4 (Fall 1993).

114. Spittlers and Bailey, *Clockmakers and watchmakers of America by name and by place*, 2011.

Addendum

Three years have gone by since this project was first submitted for publication. With the forward momentum of time, it is inevitable that stray bits of information, occasionally in the form of clocks, continue to come to light.

As we have seen, the clock makers' private and professional paths often intersected with the national questions and controversies of their day (e.g., those involving politics, slavery, temperance, transportation, interstate commerce, the lack of a national currency, patent law, debts, and the harsh treatment of insolvents). It is perhaps not surprising that some of the clock makers also engaged in political activism on issues related to the financial aspects of manufacturing. For example,

on June 26, 1834, a number of "Citizens of Hartford County," CT, entered a petition before the 23rd session of the United States Congress in favor of reauthorizing the United States Bank, stating that as a result of the recent unfriendly attitude on the part of the new Administration under President Andrew Jackson toward the Bank, "...business [has been] severely embarrassed, [and] prosperity arrested," to the extent that many branches of manufacturing, farming, and mercantile trades had been suspended. The petitioners expressed their unequivocal support for the Bank, which was then the subject of a high degree of nationwide public mistrust. In addition to the names of a number of

Figure 106A. Courtesy of Jim and Traci Stehlik.

Figure 106B. Courtesy of Jim and Traci Stehlik.

Bristol, CT, clock makers, the familiar names of a few individuals associated with clock-making at Farmington and Unionville villages are found among the signers at the bottom of the document: Timothy Cowles, E. Seymour, Cornelius R. Williams, S.E. Curtiss, Elisha Tolles, Frederick P. Hall, A.F. Williams, Martin Cowles, and George Hills. (See: *House Documents*, 23rd Congress, 1st session, 1833–1834, Vol. 6, pp. 508 et seq.) Although in the short term the clock makers lost the argument, its wisdom came to be better appreciated with time.

According to the *Proceedings of the State Convention of the Whig Young Men of Connecticut* (1840), held at Hartford on February 26, 1840, George Barber and William Bradley appeared among the assembled delegates, along with a number of their clock-making counterparts from Bristol.

At that time, convention attendees expressed their firm support for sound currency as well as farmers' and mechanics' rights.

A George Barber, Unionville, CT, shelf clock, with a 30-hour wooden movement by Bristol maker Ephraim Downs (perhaps original to its case), recently turned up on an Internet auction. Also worthy of note is the exciting discovery of a Seymour, Williams & Porter shelf clock in an unusual, true hollow column and cornice-style case (Figures 106A–06D). Its 30-hour wooden movement, Type 1.117 (1980, and unpublished update), shown in Figure 106D, was produced by one of the Terry family firms. Having no case bottom under the columns, when the clock was wound down the weights rested on the mantel or shelf beneath.

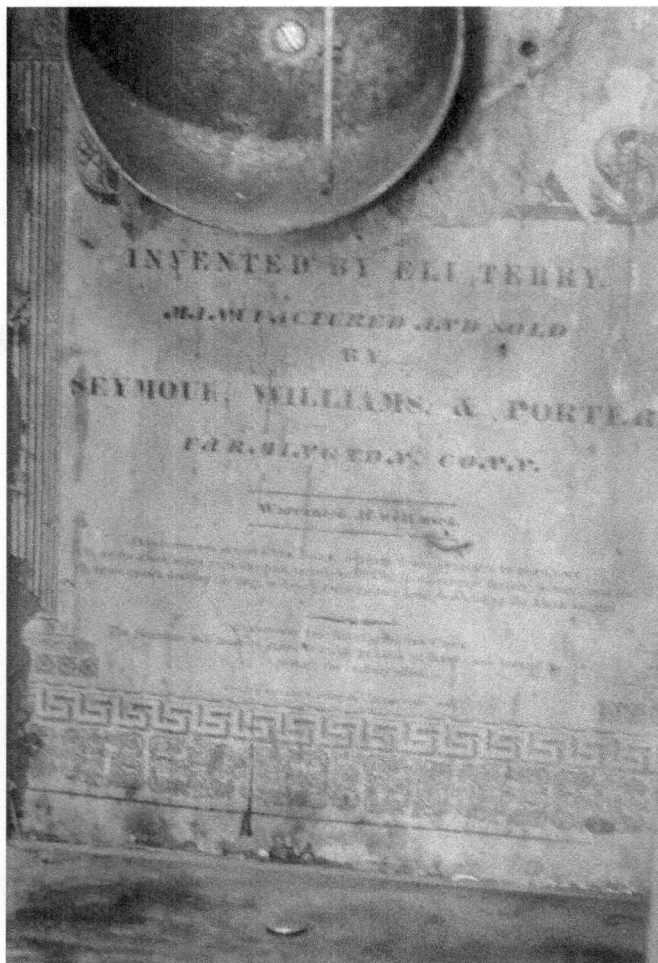

Figure 106C. Courtesy of Jim and Traci Stehlik.

Figure 106D. Courtesy of Jim and Traci Stehlik.

Index

Humphrey, Montgomery, 150
Humphrey, Ralph, 150
Humphrey, Russell, 145, 147, 150-152, 154, 157, 163, 169
Hunt, Almira, 56
Hunt, John., 84, 87, 90, 107, 110, 153, 174
Hunt, Orra, 56
Huntington, William, 46, 47
Hurlburt & Williams, 147, 175
Hurlbut, Joseph, 8, 12, 13, 14, 17, 18, 20, 34, 35, 54, 148, 152, 175

Illinois River, 114
Illinois Supreme Court, 118-121, 136
Isthmus of Panama, 78
Ives, C. & L.C., 4, 52, 80, 90
Ives, Chauncey, 4,
Ives, Joseph, 7, 174
Ives, Mainers, 144

Jacobs, 123
Jacobs, Charles C., 101, 123
Jefferson Co., AL, 77
Jerome, C.[hauncey] & N.[oble], 172
Jerome, Chauncey, 6, 41, 110, 114, 132, 159, 171-172
Jerome, Noble, 114, 159
Jeromes & Darrow, 7, 8, 110, 141
Johnson, Daniel B., 56, 57, 109, 164, 170, 171
Jones & Frisbie, 150
Jones & Frisbie, clock illustrated, 149
Jones, Henry, 150
Jones, Samuel, 100
Judd, Samuel E., 84, 85

Kaskaskia, Randolph County, IL, 75, 78
Keyes, David A., 87, 109, 161, 167
Keyes Street [Unionville], 89, 145
Kilbourn & Darrow, 158, 167-172, 174
Kilbourn & Darrow, clock illustrated, 170
Kilbourn & Humphrey, 170
Kilbourn, Betsey (Wright), 169
Kilbourn, George, 143-145, 151, 152, 165, 167-172
Kilbourn, Lucy (Peck), 144, 169
Kilbourn, Norman, 144, 169
King, Rufus, 46, 47
Kittinger, D.M.., 127, 135

Lamborn [Attorney General], 120
Langdon, Charles I., 65, 69

Langdon, Dwight, 101
Langdon, Polly, 56
Langdon, Solomon, 39, 40, 41, 42, 46, 61, 62, 63, 64, 67, 74, 76, 89, 145, 160, 162, 171, 173
Latter Day Saints (Mormons), 123
Lee & Butler, 54
Lewis & Brown, 46, 47, 48
Lewis, Henry, 57
Lewis, McKee & Co., 54
Lewis, Asahel H., 152
Lewis, Thomas I., 46, 47
Lewis, Timothy C., 76, 80
Libby, W. & Co. [Alton, IL], 123
Lincoln, Abraham, 78, 114, 115, 117-118, 121
Litchfield [CT], 34, 148, 169, 171, 173
Litchfield Twp., Herkimer Co. [NY], 34
Little, Josiah, 135
Logan, Stephen T., 118-120
London [England], 172
Long, Benjamin F., 126
Long, Julius H., 119
Long Island, NY, 140
Lovejoy, Elijah P., 114
Lovely Street, 163, 170
Lowrey, Romeo, 2
Ludington, 54
Lusk, John T., 118
Lyman, Mary Ann (Ives), 144
Lyman, William, 144

Macon, GA, 48, 172
Mann, Charles, 74
Manning, John, 118
Manross, Elisha, 7, 28, 95
Manross, Wilcox & Co., 95
Markland, Bradley & Co., 89
Markland, Susan, 89
Markland, William, 89
Markland, William Jr., 89
Marsh, Caroline (Gilbert), 6,
Marsh & Co. (see Marsh, George & Co.)
Marsh, George, 6-24, 34, 46, 85, 96, 110, 123, 132, 133, 138, 148, 160, 161, 162, 173, 174
Marsh, George & Co., 1, 2-5, 6-24, 28, 80, 83, 103, 132, 135, 136
Marsh, George & Co., clocks, illustrated, 16-22
Marsh, Gilbert & Co., 1, 4-5, 6-24, 28, 51, 52, 80, 83, 103, 160
Marsh, Gilbert & Co., clocks, illustrated, 8, 9-16